AF335105

piori I.
BIAK I.
Korida
Warsa
Wardo
Korim
Biak
Awai I.
YAPEN I.
Pom
Yobi
Samberbaba
Ansas
Serui
Amboi I.
Sawai
D'Urvile Cape
Mataboor
Apauwor
Rambai
Sarmi
Kasonaweja
Andusu
Armopa
Demta
Depapre
JAYAPURA
Tarawasi
MUSEUM
RAWASIH BAY
Waren R.
Mamberamo R.
Dabra
Ubrap
Abepura
Napanwainan
Serua
Merem
Tariku R.
Arso
Nabire
Wanggar
Wandai
Bokondini
Lereh
Tomasih
Wama R.
Beoga
Taritatu R.
Panjai L.
Mulia
Ubrub
3370
Mapia
Enarotali
Illaga
Tiom
Wamena
maniapuka Mt.
Tigi
5030
4750
Tigi L.
Jaya Crest
Trikora Crest
4595
Yamin Crest 4700
Tembagapura
JAYAWIJAYA MT.
Mandala Crest
Uta
Kapare R.
Otakwa
Kokenau
Amampare
Akimengan R.
Assued R.
Yapero
Flamingo Bay
Agats
Kaima
Waropko
Mindiptana
Kampung R.
Tanahmerah
Gantentiri
Digul R.
Wam L.
Keisak
Yondom
Muting
Bupul
Keira
Kimaan
YOS SUDARSO I.
Okaba
Kombe
Saban
Merauke
Tamerou
Kamulu
AN JAYA
96
192 Km

IRIAN JAYA - PAPUA

A TIMELESS DOMAIN

IRIAN JAYA - PAPUA

A TIMELESS DOMAIN

Photographs & Text By
Julie Campbell

Graham Brash

Singapore

Contents

I dedicate this book to —
Mal, Suzette and Yvette,
and to all my Dani friends.

Acknowledgements

The author and publisher thank BP for its most generous support in the publication of this record of the people, flora and fauna of Irian Jaya - Papua.

This easternmost region of Indonesian is a "Timeless Domain" in the sense that for centuries tribal beliefs and practices have not changed; plants have thrived undiscovered; mountains, rivers and forests have remained unexplored. The next decade will probably see more changes than in the last five hundred years, and this book will then be an invaluable record of things as they were.

BP has established itself as a worldwide environmental leader and a company committed to working sensitively and responsibly with their local communities. As the Tangguh development in Irian Jaya - Papua proceeds, and the company begins its long-term relationship with peoples of the Province, the challenges facing BP to sustain and extend this leadership position will be formidable. They deserve to succeed.

Reaching for the future, yet cherishing the past, BP will contribute towards the growth as well as the heritage of Indonesia.

Introduction

Since the first edition of this book my work has brought me to the lesser-known regions of the Berau Gulf, Bintuni Bay and the Bomberia Peninsula which are located in the southwestern part of Irian Jaya. Though the main towns of Sorong, Manokwari and Fakfak are well known, the villages and people that live in the remote places between these towns lived a life little disturbed by the outside world until the latest wave of explorers seeking oil and gas arrived.

Logging, plantations, shrimp harvesting, and transmigration development has changed the otherwise dense forest regions on both the north and south shores of Bintuni Bay and the Berau Gulf.

Historically the people of the north shore were nomads who lived in temporary huts made of forest materials balanced on stilts rising twenty feet or more above the ground.

Hunters and gatherers, they moved through the dense forests of nipah and sago palms having little or no contact with outside influence.

Following the arrival of the missionaries in the early 20th century, the nomadic tribes resettled in nucleated villages on the main riverbanks leading to Bintuni Bay.

Little was known of the area until the mid 1930's when Dutch geologists arrived and started to explore for oil. Some of today's villagers tell how their grandparents worked for the oil crews when they were surveying the area.

Village life on the Bay was little disturbed until World War II brought a Japanese invasion force to the southern shores of Bintuni Bay, holding many of the villagers captive and forcing them to build airstrips. By the end of the war, most of the villagers had returned to their simple way of life as hunter gatherers and subsistence farmers.

It is from Bintuni Bay that I begin my journey in this revised edition of ***Irian Jaya: A Timeless Domain.***

Opposite: From my helicopter seat I see below me on the southern shore of Berau Bay, the small village of Tanah Merah ringed with coconut palms and fruit trees.

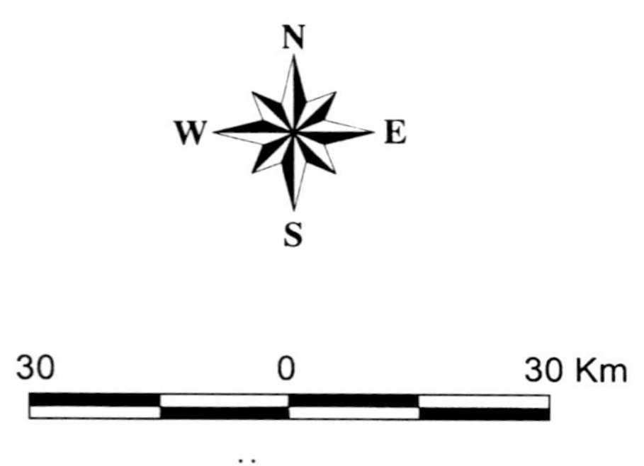

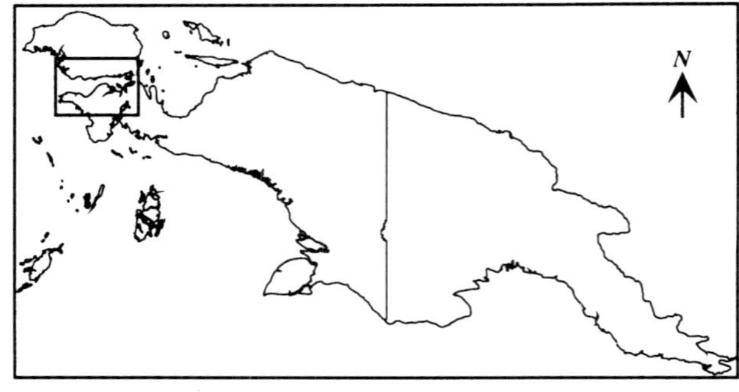

Berau Bay Region

Berau-Bintuni Bay

Berau-Bintuni Bay, fringed in the north by extensive nipah and sago palms, nearly cuts in two the Bird's Head region of kartz and limestone outcrops. Here at the most eastern part of the bay is one of the largest mangrove swamps in the world, stretching back to a forest of tropical hardwoods, wild breadfruit, palms, pandanas, colourful creepers, orchids and medicinal plants.

Besides the Bird of Paradise, once prized for its plumes at the courts of Turkey and Nepal and later in the fashion world of Paris, the swamps are home to hundreds of colourful parrots, kingfishers, cockatoos, lories, hornbills and the Victoria crowned pigeon, the world's largest and most beautiful, with a fern-like crown of white-tipped blue feathers.

The forest hides deer imported by the Dutch, cassowaries, wild pig, tree kangaroos, the shy bug-eyed gentle cuscus and other marsupials. Winding rivers harbour treacherous crocodiles, turtles, snakes, lizards and other amphibians.

Evening brings out hundreds of moths, their wings decorated with geometric designs. To name but a few of the area's unique dwellers, there are sago beetles the length of a small hand, with stick-like antennae and a bite said to be deadly; rhinoceros beetles which spit when disturbed; praying mantis nodding their heads at other winged creatures such as the puffed-out crickets.

The Bird's Head region was first mentioned in a 14th century poem. In the next several centuries the region gained notice for its trade in Chinese porcelain, cloth and metal tools in return for massoe bark, used for medicinal purposes by the Javanese, dried sea cucumbers, tortoise shell and exotic bird skins. The Sultan of Tidore claimed the area for hunting and slave trading. Portuguese settlers left their mark too in the cannons still found occasionally in some of the coastal villages, where they are revered as a status symbol and sometimes used as a wedding exchange gift. Later came the Dutch, British and Germans to explore, trade and govern.

In 1942 during World War II the Japanese established airstrips in the southern part of Bintuni Bay and the Berau Gulf, where for over three years they dominated the villagers. Today life moves forward under the protection of the Indonesian flag.

Days without Names

Over three hours is a long time to wait at any airport, but then they are not usually like the one I am in today, which has nothing to tempt the passenger in decor, comfort or coolness. Jeffman Airport is an island surrounded by azure water and fish traps, 20 km east of Sorong, once a bustling oil centre at the western tip of the Bird's Head region.

At 6.30 am the sun bounds over the horizon, smacking one's face with its heat. The sea, like a glazed eye, lies unblinking, reflecting the mooring sticks for fragile outriggers of the local fishermen. The breakfast menu at the dockside café offers dried fish, cold hard-boiled eggs, rice, coke and water - not exactly my 'cup of tea' after a five hour flight plus a two hour time jump from Jakarta.

A shady alley between the cafe and an outhouse brings cool relief in the whisper of a breeze. Red and blue plastic chairs totter on the uneven pavement, welcoming a weary bottom recently released from a narrow plane seat. Dumping my computer, camera bag and back-pack bulging with oranges, water, books, writing materials, files and the like (I always travel light), I plonk my sweaty body down and ruffle in my bag for the magnetic scrabble game. Women holding limpid-eyed babies on laps and at the breast watch with coy eyes as I unstick the scrabble letters. Soon a circle of dark-skinned coiled-haired men drape themselves on chairs nearby. "Are you playing scrabble?" one asks. Completely taken aback at finding a fellow player, I answer, "Yes. Do you want to play?" Laughingly the men tease us as he plays in Indonesian words and I follow, my spelling leaving much to be desired.

As we play the heat shimmers on the wooden pier now crowded with people spilling out of rickety boats from Sorong, passengers bound for Jakarta, Bintuni, Ambon and other exotic places. Luggage topples from shoulders and bumps against bony hairless legs whose owners drag whining children by the hand, all jostling to catch a plane. The heady smell of durian (a fruit of acquired taste and 'scent'), the rancid smell of dried fish and the acrid pungency of frying chillies flash nostalgic memories of other docks in Irian.

Pools of black asphalt oozing from the runway lie shimmering in my path as I head for the helicopter. Wearing lifejackets and earmuffs like green grasshoppers' eyes we await lift-off while rivulets form on my body in the 90 degree heat in the cabin. All is refreshed as with a great drone we take to the skies, and I am filled with the exhilaration of returning to the Irian I have come to feel such a part of over the past ten years.

Below me small islands resting on a grey metal sea give way to forest cob-webbed with early morning mist. Cockatoos rise like handkerchiefs fluttering in a breeze and from past experience I know they are screeching their displeasure at the enormous bird invading their territory.

Nose glued to window I watch an unfolding landscape of wide brown rivers snaking their way across the endless green of the lowlands. Here and there lies the skeleton of a forest, its parched skin now ready for planting. The further we fly the more I feel the remoteness of the world beneath us. Occasional hamlets sprout unexpectedly out of the green forest shelter, near swamps mirroring light.

Nearly two hours out of Jeffman I see for the first time the village of Weriagar, on the shore of Bintuni Bay and bordered on one side by a wide brown river. The seismic camp just beyond will be home for the next two weeks. Making a low pass we land, and I am welcomed to what is kindly known as Club Mud - and a busy life of nameless days.

That evening I watched hornbills thumping their black bodies across a rose tinged sky. These heavy birds with large eyes and enormous beaks always flew in pairs or more and followed the same flight path every night. Cockatoos shrieked at each other as they flew to trees near our camp, their noise adding to the deafening rasps of the hundreds of flying foxes waking in the trees on the riverbank.

Shadows lengthened as the river flowed swiftly along, carrying a few canoes laden with forest supplies and villagers on their way to Weriagar. Waving, we made our first contact. Tomorrow I would meet them.

The Village in Question

Down river from the base camp lies the village of Weriagar, home to over 800 inhabitants subsisting at the mercy of the tides which alternately flood or leave their land parched in the stifling heat. A blend of Muslims and Christians, they eke out a living fishing, hunting and gathering sago from the palms in the nearby forest. A small income is derived from selling processed sago starch, along with dried deer meat and fish, to other villagers or to the local men working at the seismic camp.

The morning of my visit was wet and blustery as we motored downriver, the wash from our boat splashing over the sides. Low dark clouds hung in the sky. It was hard to believe how cool one could be so close to the equator. Usually by mid-morning the land shimmers in 100 degree heat and stifling humidity.

As luck would have it the tide was rushing out, leaving the rickety dock drunkenly exposed in the mud. "Salamat datang! Apa kabar?" (Welcome! How are you?) cheery voices called out. "Kabar baik" (I'm fine). Eager hands dragged me up the fragile dock steps while others pushed from behind. We were about eight feet below the dock, and heavy boots and rain-gear made a ladylike ascent impossible. I had arrived!

Slipping and sliding I made my way to Kepala Desa's (Headman Desa's) home where many villagers had gathered for my visit. After several welcoming speeches and my less eloquent reply an elderly woman pulled me from my seat, tied a skirt of sago fronds around my waist and hung a necklace of shells around my neck. Then she took me to meet the other women. For the rest of the day I was chaperoned by a group of ragamuffin children gleefully hopping around me as I toured the village.

That evening we stayed for the festivities arranged by members of the Bau and Hindom families, a night which would turn back the clock several centuries. Assembled on the floor of the hut were six men, two holding a large brass gong while the others had small drums and stick instruments. As the musicians played we formed a large circle around them, stomping our feet and waving our hands. "Biak, biak, Ibu Jules, bisa menari" (Good, Mrs. Julie, you can dance), the women urged me on. Suddenly a man dressed only in a loin cloth and a head-dress of cassowary feathers leaped into the centre and began imitating the steps of the bird. Others took his lead and soon the room was pulsating with images of kangaroos and birds. The scene was a replay of the dreamtime dances of the aborigines in Australia.

While the floorboards of the hut vibrated I stepped outside into the clear night. It was a magical moment — overhead a star-filled heaven, darkness all around except where an enormous flare from the test well turned bushes and trees into eerie shapes. There I stood a witness to ancient ritual and modern technology.

Daily Bread

Over the months I became close to the village women who shared with me the joys and sorrows of their everyday lives. Together we went fishing, visiting their forest gardens, administering aid during an outbreak of measles and assisting with a nutritional programme funded by my host company. Whatever the weather, whether pouring wet or steaming hot, my visits were always welcome.

Village life was a kaleidoscope of events. Three little girls struggling with a dead river iguana. A young boy with his catch of fish suspended from a pole over his shoulder. Very young children lugging full water bottles from the water tanks to their homes. A group of young boys tossing sticks into a hole. Mothers, sleeves up, washing clothes at the river bank while their offspring hurled themselves into the swiftly running waters. Voices echoing back and forth between the men as they paddled their heavily laden canoes.

One day in particular stands out for me. Kepala Desa's wife and other women elders had invited me to photograph the preparation of *sagu* (sago) cakes and *peda*, a glue-like substance eaten with fish at most meals.

The kitchen built at the back of each hut is where children play and women gather to visit and cook. A long hearth of baked earth, over which is placed an ironwood rack for pots, takes up most of the back of the room. Buckets, bowls and various paraphernalia cluttered the floorboards. "Selamat, Ibu Jules," the women called. "Bisa duduk disini." (Greetings, Mrs. Julie. Sit here.) Room was made for me amongst legs, puppies and babies.

Sitting cross-legged on the floor, a woman removed a heap of moist sago flour from a container and put it on a woven tray which she tossed up and down to sieve the particles fluttering like moths in the sunlight as they fell back into the tray. An elder ibu, her face glowing with firelight, removed the cement baking moulds for the sago from the coals. Lowering them to the floor she dribbled sago flour into them. Other hands quickly placed a sheet of metal covered with fiery embers on top of the mould. For a few moments the babble subsided around us while the sago cakes baked.

Removing the metal and blackened coals, Ibu Desa laid the moulds on their side and taking two sticks, she gingerly withdrew the baked anaemic looking sago slabs. Handing me two sticks, she said, "Ibu Jules bisa ambil ini." (You remove one.) Trying my hand was not as easy as it looked. Sticks, fingers and heat did not dance to the same ballet as the ibu's. "Ibu Jules, tidak itu! Panas, panas! Wahai!" (No! That's hot! Oh my!) as I accidentally burnt my fingers retrieving a slab from its dark recess. "Ouch! Ma'af," (I'm sorry) I replied, trying not to appear as clumsy as I felt.

When all the slabs had been removed they were placed at the edge of the fire-pit to finish baking. After they turned an earth brown they were removed from the heat and their surfaces sanded with the skin of a stingray's tail wrapped around a piece of wood.

By that time the babble had crescendoed to a climax as children demanded food, fought with each other and unceremoniously dragged yelping mongrels out of the kitchen. Ibus fussed with trays, pots and plates as they hurried to set lunch on the floor.

Pulling off pieces of charcoal-baked fish lying glassy eyed on tin plates, we broke off pieces of hot sago cake and popped the morsel into our mouths. Hmm… not so bad! In fact, a pleasant nutty flavour tickled my palate as I ate my first lunch with the village women. Not always is sago as appetising. Usually the flour is mixed with water and cooked slowly over a fire until it resembles glue. Having tried it once, I can assure my readers it is a delicacy requiring an acquired taste. In the late afternoon I thanked the women for my visit and said my farewells.

Back at camp a friend and I took a boat and went for an evening ride up the Weriagar River. We encountered egrets taking advantage of the low tide that had exposed mud flats where juicy morsels were had for the probing. All around us the forest sparkled with life after the light afternoon shower. The scent of flower and foliage hung on the air. Flame of the Forest creepers danced in festoons of beak-like scarlet blossoms over the river bank. Dong-dong sounds of hornbills' wings disturbed the liquid silence around us.

Engines stopped, we drifted as a symphony of sounds rose from the forest and the evening sky darkened in a cloak of crimson. I remember how insignificant I felt in the solitude, the pulsating life of the swamp, while overhead myriad stars twinkled like glow-worms on a velvet cloth. The lap-lap of the river against the boat, the stillness of a world little disturbed by man. Hours and days, separated only by events, rolled into each other.

The Origin

To reach Tomu, a village situated on a tributary of the Sebyar, one has the choice of going the direct route up the Sebyar River and into the tributary to the village, or by the longer way via the Weriagar River and through the sago canals. The latter is more interesting since one travels alongside village gardens and deeper into the swamps, a place where roots, nipah palms and branches clog the waterway making travel an exciting episode.

One often meets an old 'state of the art' covered motor boat labouring along the canal, pulling a long raft of logged sago palms. Cut into metre lengths and bound with chains, they are prisoners of the launch that drags them to their destiny of processed glue that is later sent to Japan.

The canals harbour a life not seen on the wide rivers. Clusters of simple thatched shelters, home to nomadic tribesmen, litter the banks. Children frolic in the cool water while their parents process starch from felled sago palms. Troughs made from the fronds hold the orange coloured pulp which the women force with water through a fibre sieve into the trough below, where the heavy starch filters to the bottom. Fragile canoes bob up and down at their flimsy moorings while others, laden with firewood, dogs, babies, nets and the like, are paddled up and down the waterways. On other banks burnt areas herald the making of a new garden in which later will sprout root crops, beans and bananas. One stretch shows where the clearing fire got out of hand, the blackened earth holding bedraggled sago and nipah palms, their fronds hanging like brown arms from withered trunks.

Coming up against a sago raft, our boatman pulls us into the bank of grey mud to let it chug on past. We hurry as already the water is ebbing and the river to Tomu is far away.

As we approach the village of Rejosari, Ibu Camat calls out to us that we must dock at the bank to pick up a letter which she is waving at us. The docking is not so easy as glutinous mud glistens in the sun while the uneven grassy river bank beckons out of reach. Climbing up on to the roof of our small craft, three of us leap ashore. Turning around for our tall gangling co-worker, we are horrified to see him walking along the narrow edge of the boat, an accident ready to happen! The next second he slips, hangs suspended in air, and then with a plop lands on his back in the ooze. To make matters worse he tries to recover his composure, only to sink deeper into the alluvial goop. A villager leaps to his rescue, pushing and pulling him ashore. There, to add to his misery, he is promptly sluiced down with buckets of water which turn him into a soggy mess. What a first impression our poor friend has made, but the villagers standing on the bank hoot with good-natured laughter as we continue on our way, crossing into the Tomu tributary.

The Legend of the Urn

Tomu, the original village of the Sebyar tribe, is home to four clans. The major clan, the Kosepa, came from an area named Terer near the Naramasa River in south west Bintuni, On my previous visit to Tomu Pak Kadir Kosepa related the story of his people to me.

"In the area called Terer lived a man who was transformed into a snake. He put his sperm into a chalk box and sent the box to sea. It finally drifted ashore on a sandbar in Krakkak (west of Kalitami on Bintuni Bay). Several women on the beach saw the stranded box shining in the sun. One of them picked up the box and claiming it for herself, took it home to Krakkak. She ate the chalk and months afterwards became pregnant.

"In the meantime the big snake travelled from Terer to Krakkak where, through his genie, he told the people that he was looking for his wife who he believed was in the village. When the pregnant woman appeared the snake claimed her as his wife and through his genie made the villagers believe him. Putting her on his back, he swam back to Terer.

"On arrival they went inland and the snake told the woman he would cling to a tree and he wanted her to set fire to it and his body. The woman followed his orders and burned both the snake and the tree. The charred body of the snake turned into a man and the skin, falling to the ground, changed into a ceramic drinking urn and valuable goods and gold."

This legend, along with another story handed down by word of mouth, tells how after a great flood in the Bintuni Bay region the people moved away from the Terer area on great rafts, drifting northwards until they eventually escaped the flood and settled along the Sebyar River.

Dragon in Our Midst

Much of the culture of the Bintuni Bay tribes is still steeped in animistic belief. On one occasion, when the spudding of a new well was to take place, the people of Tomu came in a highly decorated canoe to the rig site. An old woman standing in the bow said, "I need to speak with the dragon about your drilling into his home. We've come to tell him that the villagers of Tomu want to hold a ceremony in his honour, but first we must feed the snake who is the owner of the tree stump on the bank near you."

The men on the rig to whom she was shouting looked at each other in disbelief. "A python in the tree? Feed it! What the hell is she talking about?" After some discussion the men decided to go and investigate.

No one could remember ever seeing or hearing of a python in the area.

Sure enough, as she had said, curled up inside a tree stump was a python, one of the largest they had ever seen. By now everyone had come ashore. The old woman, wearing a head-dress of cassowary feathers, a grass skirt tied around her waist and a necklace of teeth jangling on her old breasts, started to drum and dance around the stump. The other villagers placed dead chickens at the stump for the python to eat. They would return later, they said, for prayers and celebrations to honour the dragon.

In the afternoon when one of the drillers returned to the site to look at the python, neither the snake nor a single chicken feather was to be found. An empty hollow stump gaped back at him.

Next morning canoes laden with brightly clad villagers from Tomu again pulled into the shore close to the rig, carrying their most prized possessions: radio, cassette player, vases and clothing. Setting everything down on the covered boat dock, they formed a circle. One of the two eldest women of the group told us that she was the medium for the dragon and the other woman would translate the dragon's words. After placing 'nasi kuning' (yellow rice - a traditional ceremonial food), chicken, orange juice and food from the rig's mess on a cloth for the dragon, the two women went into a deep trance.

The medium, whose normal squeaky voice had changed into the deep voice of the dragon, said, "What have we here?"

The other woman, coming out of her trance, replied, "We bring you food and gifts, oh dragon."

"Hum! This food is OK for you people, but dragons don't particularly like your food. They eat only dragon food."

The woman translating his words then readdressed herself to the dragon. "Come and see the gifts we have brought you." The dragon in the medium's body advanced towards the gifts. Picking up each one he asked what they were for. When told that one gift made music, he said, "I want to hear this music." One of the villagers switched on the cassette player and as the music began to play the dragon rolled on his back roaring with laughter. "Why do you have such funny things?"

10

He laughed so much that all the villagers joined in, laughing and dancing around him.

While one of the rig's staff members was filming the event the dragon jumped right in front of the camera making awful faces and waving his hands at the lens. Taken aback, the would-be film maker collided into the arms of a man clad in a cassowary feather head-dress who was mimicking the cassowary's steps. Caught up in the frenzy of laughing and dancing people, they soon became lost in the throng. Soon we too were dragged into the masquerade, dancing along with the dragon who eventually accepted the gifts and gave permission for the spudding to begin.

The old medium returned to her former self and everyone packed up their goods and waving goodbye, departed, leaving us to return to yet another day on the Sebyar River.

A Small Footprint

The time has come for the seismic camp to move from the north to the south shore of Bintuni Bay. We will, however, keep in touch with the villagers whose lives over the past months we have touched but, hopefully, not changed.

As we travel down the Weriagar River there is an emptiness and loneliness in the abandoned fly camps with their activities and chatter silenced. Ladders to nowhere lean against the river banks. The framework of an abandoned shelter stands draped in emerald green creepers. Forest debris lies where small trails existed. Man had come and gone, barely leaving a trace of his footsteps in the mangrove swamps and forest. The simple village life holds changes from our visit, two different cultures living side-by-side. Now the river flows unhampered with our traffic while the morning hangs in stillness at our going.

Lest We Forget

The south shore of Bintuni Bay is an area of limestone outcrops, sandy beaches fringed with palms, villages tucked into the bends of rivers full of fish and oyster beds. To the east stretch vast tracts of mangrove forests, while inland savannah plains are treed with paper-bark eucalyptus. Trawlers ply the edges of the bay in their endless search for shrimps. Long legged egrets, heron and sacred ibis stab the mud for delicacies. Kite, sea eagle and common terns shriek in the wind as they follow the fishermen.

Here too are the old Japanese airfields with the bombed out carcasses of their fighter planes, a leaden reminder of when the villagers were held prisoners on their land and forced to labour for their Japanese taskmasters from 1942 to 1945.

In the western part of the bay, rising out of the aquamarine sea, are high and tree-crowned islands where waterfalls cascade from mountains 4,000 feet above sea level. Old nutmeg plantations, transmigration settlements and logging camps intersperse the land. Villages, baking in the sun and ringed with sea-going craft, hug the shoreline.

The aura of peacefulness that lies over the land shelters the easy and hard times of those who live there. The villagers' lives are bound to the few opportunities for work outside their villages at the shrimp factory, the logging camps, or the plantations. Those who remain in the village teach the children, raise crops in local gardens and fish the river and bay with simple nets.

Tanah Merah, a small village cradled in a horseshoe bay to the east of our camp at Saengga, holds memories of playful children whirling new frisbees, skipping ropes of rubber bands dusting the beach, women whacking at the carcasses of large fish brought home by their men, carpenters and brickmakers working into the late afternoon hours at the new church they were building with money saved over the years, and afternoons when the children and I used metal spoons to dig scuttling crabs and shrimp out of the rocky crevices.

The time when Pak Siwana presented us with a bowl of freshly picked mangoes whose thick luscious yellow flesh spilled cloying juices over our hands as we ate the fruit. The wild storm the night I stayed in the village. Waves crashing, branches of the tall mangroves creaking in the wind. Boats bobbing in the bay while we played flashlights on our driver, his oil slicker a golden orb enveloping him as he tried to secure his craft in the inky darkness.

A seven kilometre walk through the rainforest steaming in the heat, its streams swollen from the previous night's storm. Leeches clinging like thorns to one's legs, hands and clothing. The joyful buzz of insects as the sun dried their wings. The mossy smell of a new day.

Each village has its own memories. In the far eastern sector of Bintuni Bay lies Babo, a village whose history dates back before the Royal Dutch Petroleum Co. settled there in the early 20th century. A history built around the skeletons of war when Japanese invaded the area and, with the villagers as their prisoners, built an airstrip on the outskirts of the village.

I was fortunate enough that day to find the oldest inhabitant of Babo at home. Pak Kwendo, a sprightly lean man of over 80 years, lives near the old Dutch jetty in Babo. When I asked him if he remembered the events that happened in the village during the Japanese occupation, he said, "Saya tiduk bisa lupa (I cannot forget). Bisa masuk rumah saya (Please enter my home)." It was a relief to enter the coolness of his simple two-storied house where the walls were adorned with old photographs of his family. Settling himself into a wicker chair opposite me, he began to unfold the pages of history as he remembered it.

"My father was a Hokkien trader from Ichan in China. He came here over a hundred years ago and married a woman from the island of Kaio. They settled here in Babo where I was born, along with five brothers and sisters. We went to the Dutch missionary school and later I worked for the Royal Dutch Petroleum Co. We grew our own crops and took care of our animals. It was a peaceful life shared with other traders that lived in our small village.

"In the summer of 1942 many planes flew overhead. We rushed to look. Some of us were afraid and knelt in the fields to pray. After a while the planes left, but later one returned and dropped a bomb near the village, killing one woman and destroying some of our huts. Some time later boats landed on the beach and soldiers entered the village. Many of our men were taken away and others of us were put into uniform and made to work for the soldiers, building an airstrip for them."

"What did the women and children do?" I asked.

"They had to grow rice and vegetables, and work for the Japanese. We were prisoners in our own village. We had our hands tied behind us and were made to kneel in the hot sun if we did not do what they wanted. Life was very hard. Many of us were beaten and over 20 villagers were killed. One of them was my brother." Pak Kwendo paused and looked at a photograph on the wall before continuing his story.

"When the airstrip was finished many planes landed. Every day some of them left the village. One morning a plane was shot out of the sky by the Japanese. We watched it crash and later that evening a villager found a wounded man on the beach. We looked after him and hid him for a very long time.

"Soon after that 20 planes left the airstrip but only eight returned. Different planes flew near the village and dropped pieces of paper on the beach. We gathered them up, hiding them until we could safely read what was on them. The words told us that we must leave the village within two days as Allied planes would bomb it. That day some of us secretly prepared for the journey. The next night was very black as we stole down to the beach and slipped into our canoes." Here Pak Kwendo got up from his chair and acted out how they had moved in the dark night, hiding from their captors.

Returning to his chair, he continued, "We paddled our canoes very quietly for a long time until we landed at Geser Island, where we dragged the canoes up the beach and hid ourselves deep in the forest. We stayed there until we couldn't see or hear any more planes. When we returned to Babo we found big bomb holes everywhere, and on the airstrip many burnt and damaged planes. The stores where our rice and food were kept had been destroyed. We had so little left that we had to hunt in the forest and live on what we could find.

"Some time later a boat full of Australians landed on Kasori Beach. The men talked to us and gave us food before leaving to search for Japanese soldiers." By this time Pak Kwendo, his eyes aflame, was gesticulating and dancing around the room as he acted out those last few days of the Japanese surrender.

Most of the morning had gone when I finally thanked Pak Kwendo and left him with his memories. Walking through the village I could still see many bomb craters left from World War II. At the end of the village the small path continued for about 20 minutes, bringing me to the old Japanese airstrip, now used once a week for the Merpati flight to Bintuni and Manokwari.

14

Sitting shining in the sun were skeletons of small bombers with sightless eyes and empty noses where propellers once had been.

Freedom Is My Song

All journeys must come to an end, and for me one of the most moving times in my work in Bintuni Bay was the day I said goodbye to the villagers at Weriagar. A small boy gave me a beautiful bird, squat of tail, with brilliant blue wings striped with black, a deep russet coloured body, and a golden ring of feathers around neck and beak. The beak was broad and shaped like a short spear. I recognised it as the young Irianese kingfisher. Earlier I had seen a small child dragging it across the floor by a string tied to its leg.

We took the bird across Bintuni Bay to our Babo base camp because I felt that if I abandoned it to its plight it would not survive, that it would never know the power of freedom, the joy of song, the right to be wild and free.

Once at camp we put it in Sawarno's office where it perched on a makeshift rod in a corner of the room. Everyone took turns feeding it small pieces of shrimp that we pushed down its throat since it could not feed itself. It would cheep and open its beak for more, so we knew we had made a hit food-wise. It was cold in the office, however, so at night we moved it to the warmth of the generator house.

On the third day I went into the office and could not find it anywhere. The bird had flown across the room and settled on a wire leading to the computers. After that we all decided it should be released to fend for itself since it seemed able to fly, and besides, it had started to refuse the shrimp we fed it.

During the night I tossed and turned as I felt uneasy about letting the bird go free here on the south shore where everything was so unfamiliar, even though some of the locals told me they had kingfishers in the area. Waking the next morning I knew what I would do. We were due to go back to Rejosari on the north shore to meet with Ibu Camat. It would be easy to return the bird to its natural habitat at the same time since we were to go by boat, returning by chopper from the old Weriagar base camp.

Sawarno, my companion and bird-sitter, found a box in which the bird could travel with us across Bintuni Bay, a journey of one and a half hours. It was a beautiful morning. Everything shimmered after the rains. The sea, reflecting the sky, lay like a blue mirror. Near the middle of the bay dolphins accompanied us for a while, and a turtle popped its head up at our passing. The bird sat hunched on its perch inside the box, withstanding the engine's roar and the jolting journey, its second across the bay in four days.

Speeding up the Sebyar River to Ibu Camat's office, the bird seemed to take an interest in its surroundings as we drew near to the dock. After handing a letter to one of the assistants we returned down the Sebyar since the short cut to the Weriagar River through the sago canals was impossible at low tide.

As we approached the mouth of the Weriagar the bird hopped off its perch and onto the ledge of the boat. Hurriedly I retrieved it in case it came to harm. The difference in the bird was amazing as we neared the area where I learned it had been caught. It fluffed its feathers and its large rimmed eyes shone brightly as it kept turning its head. It obviously knew where it was.

Pulling slowly into the old broken down jetty at the camp, Sawarno climbed up the ladder, taking the box from me as I clambered up behind him. Holding the box gingerly, with the bird teetering on the edge of it, we walked slowly down the rickety steps into the former camp area.

When we neared a stand of young sago palms the bird suddenly flew out of the box and landed on a lower frond of the nearest palm. We stood still, a great feeling of joy and relief at the bird's apparent recovery. We waited to make sure it was safe before we retraced our steps to the bridging near the heli-pad. After a while we heard that familiar long drawn chirp which I had come to recognise. Then the kingfisher's call grew fainter as it moved further into the forest.

Palm fronds clacked in the breeze. Discarded rice grains, now re-seeded, jostled their silvery tassels in the sun. Hornbills, their great wings making the well-remembered dong-dong sound, flew overhead. Cockatoos screeched at each other as they swooped in and out of the trees, and stately egrets billed their way into the riverbank for food.

Everything was so beautiful, but there was an added magic that morning in the thrill of returning a captive creature to its natural environment.

As the helicopter whirled nearer to us, the last sound I heard from the forest was the call of the bird who so unexpectedly had shared a brief moment of its life with us.

Weriagar village, home to over eight hundred people, lies between the Weriagar river and the north shore of Berau Bay.

Opposite: The nomadic people of the northern Berau Bay area build their huts along the riverbanks. Often these shelters are used as temporary housing when the families harvest the sago from the swamp forests.

The main source of food for the tribes of Berau Bay comes from the sago palm in the form of starch. After a sago palm is felled the bark is pried off and the fibres pounded into pith. Fronds from the palm are made into a simple trough at the head of which is placed a cloth or fibre filter. The sago fibres are piled up against the filter and water is then poured over them leaching out the sago starch. Starch settled at the bottom of the trough is scooped out and packed into bales and brought back to the village. In the family cooking quarters the women pour the starch into a cement mold. a piece of metal holding hot caols is placed over the mould. After ten or fifteen minutes the dried blocks of sago are removed and balanced around the firepit to finish baking. Taking a scraper made from the tail of stingray Ibu Hindom removes the burnt pieces from the sago blocks and lunch is ready. (refer to page 20 & page 21)

In a canal dug by the sago company Sagindo Sari Lestari in Rejosari a small boat drags harvested sago palm, a process known as sago rolling to the sago factory. (refer to page 21, bottom)

This page: On an inlet of Sebyar river a fisherman casts his net hopeful of early morning catch

Opposite page: Greeting me shyly a small boy from the village of Weriagar returns with his fishing net after a successful morning of fishing.

A woman and her son from the village of Tanah Merah removes a few small fish from her net in the early morning light.

Proudly a young men holds high his night time catch for me to photograph.

I watch with interest as Pak Yuri from the village of Tanah Merah removes the fins of a small shark which he will later dry and sell to Bugis traders. The stingray will be cut into pieces and shared with other villagers.

A woman form the village of Tomu on the north shore of Berau Bay turns over the strips of fish she is drying for sale.

Moving day for a temporary village that was engaged in sago harvesting from the nearby forests.

In a dugout canoe a husband and his wife paddle home pandanus fronds and bamboo stalks. Both these items are used for building village huts.

A net handler on board a shrimp trawler owned by the shrimp factory Wimro hauls on board a catch of shrimp. Many trawlers ply the waters of Berau-Bintuni Bay day and night in their quest for shrimp. After sorting most of the catch is graded, iced and sent overseas.

The oldest Dutch home built in the 1930's still used as a private residence in the village of Babo cn the shores of south Bintuni Bay.

Paddling along a small river we disturb flying foxes hanging from a tree on
the river bank.

In the nipah swamps (nypa frutiicans) along the Seringga river a greater
egret (egretta albo) flies low across our bow.

Irian Jaya and its People

Irian Jaya's neolithic culture, an isolated remnant of an era long past, is rapidly vanishing as the outside world, in the form of missionaries, officials and tourists, creeps into its once hidden domain.

Curiosity prompted me to make my first trip into Irian Jaya, the Indonesian half of the great island of New Guinea. Captivated by what I saw, over the last ten years I have made innumerable visits to various parts of the country, photographing a people in the process of discarding their cultural heritage as they come to terms with the modern world.

Although there are over 100 distinct tribes scattered over Irian Jaya's coastal, swamp and highland areas, in this book I have focused on only four groups: the Dani of the Baliem Valley, the Jale of the highlands, the nomadic people of the Brazza River region and the Asmat of the lowland swamps. These four groups are broadly representative of the traditions found among other tribes.

Nestled in the Baliem Valley about 1,700 metres above sea level and surrounded by a mountain wall, the Dani, until visitors brought them modern shovels and axes, continued to use the Neolithic (stone) tools of their ancestors.

The Jale, hunter-gatherers, eke out a subsistence living in the highland forests. Unknown to the West until 30 years ago, they live mainly on sweet potatoes cultivated on the mountain slopes near their small, remote villages.

Opposite: Above the Baliem Valley a Dani village, with its sweet potato gardens etching the hill side, creates within the protective stone walls a warm haven for the villagers.

The reclusive nomadic tribes live in tree houses, hunting with their bows and arrows in the dense rainforests of the lowlands. They offer the strongest resistance to the outside world.

The Asmat share their swamp and coastal world with myriad animal, plant and ancestor spirits. In the West they have gained fame as producers of some of the finest and most sought-after primitive wood carvings on earth.

More than a century ago the British naturalist, Alfred Russel Wallace explored Irian Jaya during an eight-year sojourn in the Malay Archipelago in search of rare birds of paradise and other exotic species. He also observed the people of the islands and with uncanny perception, concluded that the Malay races of Sumatra, Java and Bali seemed well adapted to survive a colonial onslaught culturally intact. Not so the people of New Guinea. "If the tide of colonisation should be turned to New Guinea," he wrote, "there can be little doubt of the early extinction of the Papuan race." He added, "A warlike and energetic people who will not submit to national slavery or to domestic servitude must disappear before the white man as surely as do the wolf and the tiger." (*The Malay Archipelago* by Alfred Russel Wallace, 1869, republished by Graham Brash, Singapore, 1983.)

Wallace could not have envisioned that tourists, missionaries, officials, and multinational mining and logging companies would be the real threat to Irian customs and traditions. This book seeks to capture, in text and photographs, the secrets of a culture soon to be lost.

In the 1960's U-mue, meaning 'Anxious One', became immortalised in Peter Matthiessen's book, *Under The Mountain Wall*. At that time he held the title of *kain* or chief of the Wilil clan in the Baliem Valley. His easy humour and vanity earned him his reputation of being a 'ladies' man. Known as Chief Wali, he lives in the village of Opigama along with his three wives, his sons and daughters, his father and his cherished granddaughter, Hesagi.

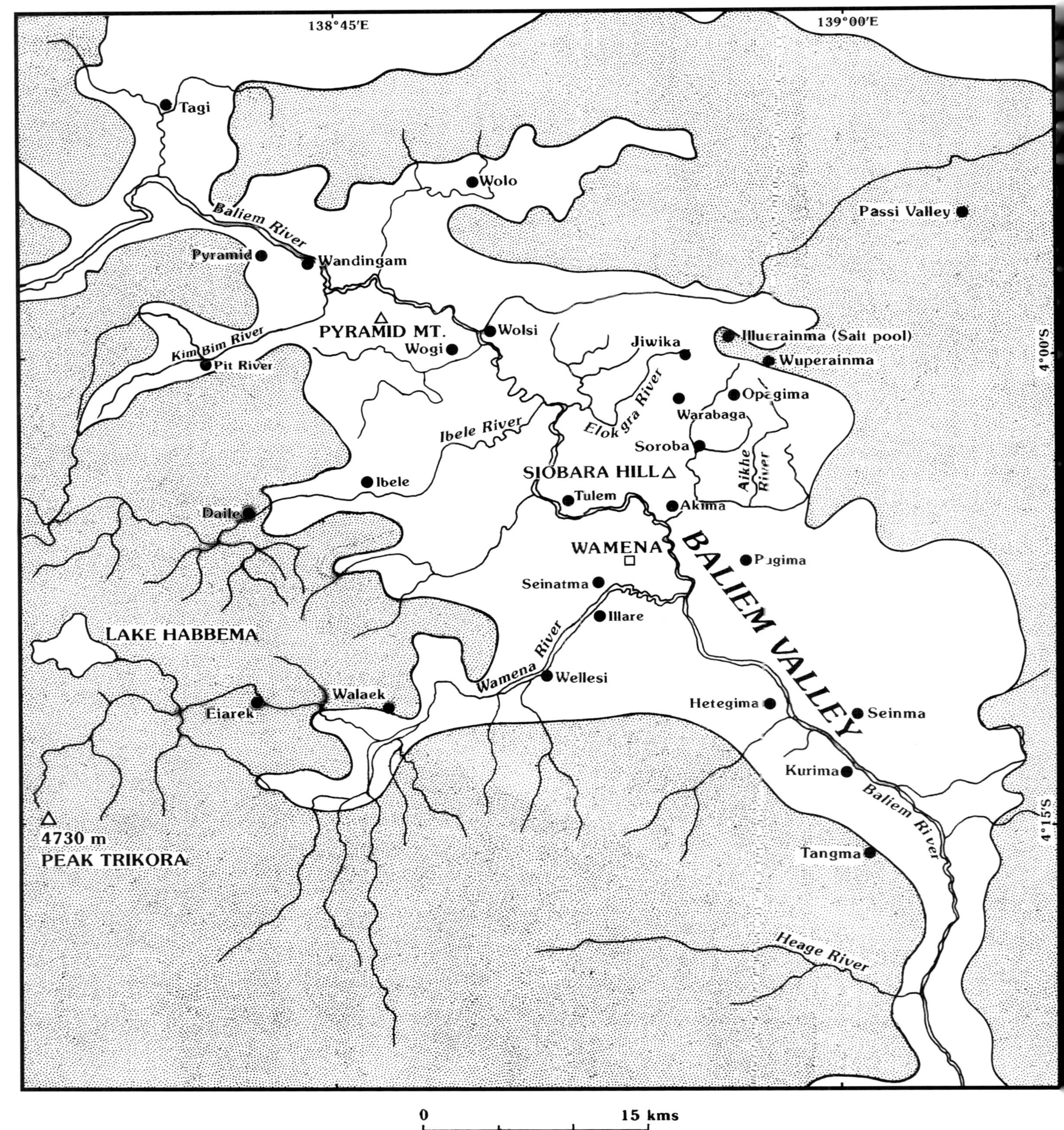

Baliem Valley Region

The Dani of the Baliem Valley

The Place and Its People

Far from the western world, on a morning in 1938, the clouds covering the Snow Mountains of central Irian Jaya parted, unveiling to an American exploration team a secret valley — a valley with intricately designed irrigation canals, meticulously terraced mountain slopes and neat gardens enclosed within stone walls.

The National Geographic, in its March issue of 1941, wrote of the discovery; yet it was not until 1945, when missionaries made contact with some of the estimated 100,000 tribesmen, that the outside world was made aware of the valley's inhabitants.

The Baliem Valley lies approximately 1,700 metres above sea level. Running west to east of the surrounding forest-clad mountains, it is 16 kilometres wide and 60 kilometres long. The Baliem River threads its way across the plain to the Baliem Gorge where it drops 1,500 metres, forming spectacular waterfalls before continuing its journey to the alluvial plain in the south, and on to the Arafura Sea.

The inhabitants of the Baliem Valley are a Neolithic race which has populated the area for hundreds of years. The origin of the Kurelu people who speak the Dani language is unknown as they have no written language. Anthropologically, they are called Papuan or non-Austronesian speakers. A stone-age people, the warrior-farmers used wood, flint and stone for their weapons and tools, until the late 60's when steel was introduced to the tribesmen.

Their own version of their origins is mystic and delightful: at the beginning of time God created a man and a woman from one large stone. Because the stone was white God called the man *Tabhe* (white skin); the woman he called *Hesage* (beauty).

God left them in the jungle where they fed on the wild fruit, The sun burned their bodies so they covered themselves with mud. At night they took refuge in a cave God had made for them (thought to he near Abulakama).

In this cave their children were born, black-skinned to protect them from the strong rays of the sun. When they grew up the children left the cave and went into the land where they built their own houses with roofs shaped like the roof of their parents' home.

When their children no longer needed them, Hesage turned her body into burning grass while Tabhe transformed himself into brown leaves whereupon God threw the leaves onto the fire and the burning souls of Tabhe and Hesage entered into the stones inside the cave. The children returned to find their parents gone. Some of the children collected the stones, which they regarded as *wusa* (sacred), and took them back to their own homes. The other children remained in the cave, which they are believed to inhabit still.

To this day there are *wusa* stones in all Dani homes and sacred places, and it is forbidden by tribal law — and accepted by Indonesian law — to remove even a fragment of these stones.

The Dani rub the stones with pig fat to keep the spirits calm. Sometimes the spirit within a stone loses its strength and then the stone is considered not sacred any more and can he given in exchange or even sold at funerals.

Scattered throughout the valley is a network of small foot trails leading to enclosed nuclear villages of mushroom-shaped thatched huts. White tendrils of smoke from the huts entwine in the morning mists. Nights are cold; the wind screams through the *Pintu Anging*, the Door of the Wind at the eastern end of the valley, while the Dani huddles beside the primaeval fire within the cocoon of his hut.

The tribal people still live with their age-old traditions. Disputes are dealt with in tribal courts, and carefully regulated mock wars are enacted in the surrounding countryside until those who have been wronged are appeased according to thc ruling of the court. *Wusa* (sacred objects and places which afford protection against evil spirits) abound throughout the valley in stones, trees and caves. Ancestral ghosts are taken into consideration in the Dani's daily life and the great pig feasts are observed according to ancient rites.

The men's traditional dress of penis gourd, and the carrying nets and fibre skirts of the women are still worn. Sweet potato gardens and irrigation canals prepared in much the same way as those seen from the air in 1938 are in some areas still tilled with the traditional ironwood digging sticks.

Marriages are held in the valley once every four or five years, at the decision of the great *kain*. Sexual abstinence is observed during pregnancy and the three or four years after childbirth when the mother is still nursing her young child. During that time the husband, according to his wealth in pigs, is able to take another wife.

The *kain* is the undisputed ruler of his village. He lives in the largest hut, known as the *pilamo*, with its doorway facing the entrance to the village across the silimo (courtyard). Here he and the other men eat, sleep and gossip. The wooden-planked, thatched hut is warm and cosy, with smoke from the smouldering fire blackening the interior of the hut and asphyxiating the persistent mosquito. At the back of the room sacred stones and ancient weapons lie in bark-wrapped packages. A low sleeping loft is reached by a simple wooden ladder to the right of the fireplace. The village ghosts keep watch from the passageway built between the inner and outer walls of the hut.

Beauty is in the eye of the beholder, particularly in the eyes of the male community. The Dani man spends hours with a male friend, plucking unwanted body hair with wooden tweezers of *araucaria* twigs and styling his beard. The impenetrable mat of corkscrew hair is rubbed with pig grease and tweaked into miniature coils which are adorned with feathers and flowers. His body glows with the same blackened grease. Around his neck he wears a *wa'imo*, a bib of woven bark sewn with rows of small snail shells, which protects him from the eyes of marauding spirits. *Teken* (bracelets) of dog fur or bracken encircle his upper arms. Over his penis he wears the *holim* (penis sheath), an elongated gourd, and as a symbol of the prestige and authority befitting his rank, a cuscus tail dangles from the extra-long *holim* of a chief.

The women, strong and buxom in youth or saucy-smiled in child-hood, sway like rumba dancers in their *thali* (grass skirts). When she gets married a young woman exchanges her *thali* for a cage-confining skirt of white, red and yellow orchid fibre which is braided by the males of the family and wound tightly around her hips by the village women.

She toils from dawn to dusk, gathering firewood and edible greens, collecting salt from a pool high above the valley floor, rearing pigs and children (in order of priority), and tending the sweet potato gardens. Her basic diet is the sweet potato, supplemented with tidbits such as cucumber, vine leaves, peanuts and ginger root. Only at marriage ceremonies can women partake of cooked pig meat.

Her lullabies are for her infant; her tears for her lonely birthing and, as a child, at the traditional finger-joint amputation ceremony performed at the funeral of close relatives. As an expression of grief for the dead, the women roll in the river mud, caking their bodies in sorrow.

Steeped in ritual, the Dani salutation takes different forms. *Niak* is reserved for the greeting between two males; *lauk* is used between females or by a man greeting a woman. Throughout their daily lives, the worlds of man and woman are separate. The warrior-farmer's or chief's hut is forbidden to women, girls beyond puberty and the suckling child. At tribal gatherings the women sit separately from the men and eat from their own steam-pit. The meat is prepared and eaten by the men and boys only. Only at marriage feasts is the meat shared with the women, and the mothers of the brides are given the honour of receiving the pigs' tails, normally reserved for the chief of the village.

Of late the Dani warrior has no tribal boundary to defend, no wife-stealing to appease in battle, no watch tower to build. His arrows and spears are for the hunt alone and when not hunting, his time is now spent farming, fencing, building huts, as well as learning new ways and methods. The highlight of his days is now the tribal pig feast at a ritual gathering.

With the winds of the 20th century sweeping through their *Pintu Anging* into their secluded valley, the young Dani are reaching out to new horizons beyond the range of the once protective mountain walls. The world they know may soon be lost.

Opposite: A Dani serenades his grandson with a tune on his *bigon* (mouth harp).

42

A Woman's Way of Life

I am a Dani woman, born in the Baliem Valley which is ringed by lofty mountain peaks in the highlands of Irian Jaya.

The early mornings in the valley are shrouded in mists that give way to the heat of the sun as it rises over the tree-clad mountains, warming the land of my dark-skinned people.

Often in the evenings the winds sweep into the valley, cooling the land and sending the Dani home to the warmth of our round huts, safe from the dark, spirit-filled night.

As a baby I was put on a bed of ferns and placed inside my mother's *noken* (net carrying-bag) and carried on her back wherever she went. At times, being bumped up and down was very uncomfortable and I would cry out in protest. Then my mother would sing me to sleep with a Dani lullaby.

When I grew older I wore a *thali* (grass skirt) and a small *noken* in which I carried peanuts and sweet potatoes, just like my mother.

My early childhood days were spent with my mother and my father's other wives, whom I also called "mother". we all lived together in a compound which housed the *pilamo* (chief's hut), *ebeai* (a small hut for women), *huni* (cooking hut) and *wamai* (pigs' hut). All these were surrounded by a high stockade with two entrances, one for the people and one for the pigs.

Here in this compound I learned how to finger-weave and care for the pigs. I played with the other children and sang Dani songs around our small hearth in the cooking hut.

We slept in our warm, round hut which my father had made from saplings and planks of wood, topped by a roof thatched with long grass and secured by vines.

As it is taboo for Dani women to enter the men's hut, my father would visit us sometimes, and eat with us around our hearth in the cooking hut.

With each dawn came the grunting and squealing of the pigs — a signal for my grandmother to wake and let them out of their hut. She would cross the small courtyard and open the entrances of our compound. Chickens would fly across the corral and the dogs bark. In our cooking hut we would heat sweet potatoes for our morning meal. So began our day.

Each morning I would put my arms over my chest, hold my hands behind my neck and jump up and down to get warm. I would watch my mother fill the drinking-gourd, put on her *noken* and take her walking stick. Then I'd know it was time for us to go to the gardens to work.

Fields were prepared for planting by the men of the village. Their task was to cut the roots of the trees so that they would die, dig the canals and turn over the black soil in readiness for the women who would plant, weed and later harvest the sweet potatoes.

The women's work was hard: breaking up large pieces of earth, weeding, making the mounds for planting. We'd push pieces of sweet potato vine into the mounds of earth, leaving a small part exposed to the elements. In three or four months new vines would spread and cover the ground, bearing, under the soil, new potatoes to be harvested.

While the women worked in the gardens we children would play and catch fish in our carrying-nets in the canals.

I remember from my childhood when my mother's brother died. It was a time of sadness for the family but also of excitement in the village as preparations for the funeral took place.

In the morning, many Dani warriors, shiny with blackened pig's grease, arrived in the compound to exchange gifts and slaughter the pigs for the funeral feast. The men looked magnificent in their finery — headdresses of coloured birds' feathers, armbands of dog fur, and boar tusks through their noses. The village chiefs wore their *walimo* (collars of bark and shell) and at the end of some of their *holim* (penis sheaths) hung a long cuscus tail.

Many of the men had brought gifts to the funeral which they placed in front of the men's hut. I sat with all of the other women and children of the tribe outside the cooking hut weeping while the men prepared for the funeral pig feast.

The next day as soon as it was daylight several of my friends and I were taken outside of our huts to the courtyard where we were to participate in the *iki palin* (finger cutting) ceremony performed to placate the ancestral ghosts.

I watched fearfully as one of my sisters had her left arm tightly tied above the elbow with a piece of twine. To numb her hand to lessen the pain, her elbow was banged down on a rock, after which my uncle pressed her hand out on a board and quickly chopped off the top joint of the second finger with a small stone adze.

I was so scared when I heard her screams and then, before I knew it, it was time for me to give my finger joint to the ancestral ghosts. I remember the sharp pain of the adze and my mother wrapping my hand with healing leaves which she had soaked in water and binding my arm tightly with wet banana leaves and tying it into a bundle with twine. For the rest of the day I sat numb holding my throbbing hand upright to make it heal quicker.

All of the severed finger joints were put into a small net bag and hung in the kitchen hut until the next day when they would be placed on the cremation fire.

As I entered womanhood, the time came for me to take part in the great *mauwe* (communal wedding ceremony), which is held throughout the valley every four or five years.

Many moons before the gathering, the male members of the brides' families go into the forest to collect the yellow, red and white orchid fibres used in making the braid for the *jogal* (a hip-hugging skirt for married women). Many metres of fibre are woven into a braid which is wound into a huge ball and kept for the women to make later into *jogal*.

During this period, in the chiefs' huts throughout the valley, heated discussions take place and bride prices are fixed. Brides are not chosen for their beauty, but for their strength and ability to raise pigs, tend the garden and bear children.

Each bride is expected to bring a number of new *nokens* with her when she moves to her husband's home. Weaving bridal *nokens* keeps the gnarled, stumped fingers of the grandmothers — as well as the more nimble fingers of the mothers, aunts and sisters — very busy. Their fingers fly in and out of the bamboo strips holding the string in place. When the nets reach about one metre in length, the bamboo strips are removed and the edges and headband of the *nokens* are finished in the privacy of the weaver's huts.

For some time previously the women had been cutting and gathering small trees and branches, which were then carried to the compound, along with huge bundles of grasses from the swamps and ponds. The men stacked the wood in great piles outside the corrals of the villages, in preparation for the fires. Many pigs were rounded up from the countryside, while all available *nokens* were filled with sweet potatoes from the gardens.

At last the great day dawned. Before sunrise I went into the cooking hut, together with the daughters of my other mothers. It looked different that morning: dark, smoky and full of busy chattering women. A pole had been fastened across the width of the hut, at chin height. Here we could rest our arms as we stood for the making of the *jogal*.

A neatly bound ball of red, yellow and white orchid braid lay in readiness for the winding of the *jogal*. Close by, two women rolled bark fibres on their glistening thighs making string.

A piece of cloth was put around my hips, and the grass skirt I had worn throughout my childhood was removed. Then an old *noken*, made into a bundle, was placed between my legs so that the *jogal*, when finished, would fit tightly around my hips, leaving me just enough room for walking.

Two women wound many metres of bark string around my hips and down as far as my upper thigh. The coloured braid was carefully arranged over the string so that it covered it. As one weaver wound the braid over my body another woman pulled the under strings up at the sides so the *jogal* would fit tightly on the hips leaving the skirt dipped back and front.

By sunset when all of the *jogals* were finished we had the cage-like skirt removed from our bodies. It had been a long busy day so it was with much relief that we returned to our huts to sleep.

On the third day of the marriage festivities, wearing my new *jogal* and *nokens*, I walked in procession to my husband's village where I entered the courtyard and stood before my new family. As I remained standing with bowed head, the bridal gifts were laid one at a time across my back as a symbol of acceptance of the gift.

My father kept the *je* (bridal stones) and the *jarrak* ('pig counter', a woven long narrow band of bark fibres demarcated in cowrie shells, to measure the girth of a pig), as well as the pigs. I kept the *jogal* as well as the new *nokens* which, in accordance with Dani custom, I wore on my head for three months.

After this ceremony my husband and I went to the new hut he had built for me. Sitting outside the door, we shared sweet potatoes and pig meat from the wedding feast.

From that day onwards, I lived in my husband's village sharing the work load and visiting with the other women and children around my own hearth in the cooking hut.

Many months later I gave birth. Since it is taboo for a new born child to be seen or named for a full moon we lived alone in our warm round hut.

Then the morning came when we could leave the hut. I laid my little girl, Hesagie, on fresh fern leaves and placed her in a *noken*. Covering it with a beaten bark cloth, I placed the net on my back and left the hut singing a lullaby as I joined my sisters on their way to the gardens.

A Dani Pig Feast

In the furthermost corner of my mind lurk the sounds and smells of our pig feasts. Since I was born I have been witness to these ceremonial feasts, a highlight of our way of life.

Work begins days before a feast. From the surrounding countryside we women gather great bundles of long savannah grass, balancing the bundles on top of our heads, while others carry heavy loads of rushes, and thick leaves. In the forest the men chop wood stacking it for collection for the fires.

On the day of the pig feast, villagers from far and near journey along the well-worn paths to our village, calling out to us as we gather in our *nokens* the *hibiri* (sweet potatoes) and vines and leaves of the *hela* (mallow plant) needed for the feast.

Gathering up our *nokens* we return to the village leaving them in the *silimo* (courtyard) while we visit in the cooking hut.

In our absence, Pau, the brother of Siba, has prepared a new *bpakate* (steam-pit), scraping away the grey wet soil with his digging stick to make a conical pit about two metres wide and one metre deep. Untying the bundles, he layers the long grass in the pit, letting it spill over the edge, just as the great river spills over the land.

Our men build a pyre of planks and split timber, and place large stones in the centre of the pyre.

During these preparations men from the surrounding villages arrive with their pigs as gifts to be slaughtered for the feast. All the pigs are herded into the courtyard, where they run around frantically, squealing and lighting with each other.

Three men succeed in capturing one of the pigs. Carrying it by its ears and forelegs, they bring it to Pau, who is waiting outside the men's hut with his bow and bamboo arrow. Aimed from no more than a metre away, the arrow finds its mark behind the foreleg and into the lung. Pau steps forward and plunges the arrow further into the wound.

The squeals of the dying pig rend the air as the men lower it to the ground and step on it to increase the flow of blood. This one dies quickly; not so others which, after being shot, careen around the courtyard, scattering us as they race in their death throes before finally succumbing — a quivering, blood-stained, mass.

When all the pigs have been killed, the men position them on their stomachs, heads and front paws crossed facing the *hunila* (cooking hut). Then the ancestral ghosts know that we have prepared everything as to our custom.

After this rite the ears and tails of the pigs are removed, wrapped in a banana leaf and put inside the chief's hut; later for use according *wesu* (sacred objects).

By now the pyre is blazing and ready for the singeing of the pigs, which are placed on the fire. With long sticks the men scrape the bristles from the skin until it is smooth and white.

The carcasses are removed from the fire and placed on large banana leaves in front of the chief's hut, where they are butchered with bamboo knives. Starting from each side of the mouth, the men cut down past the stomach and between the legs to the anus. Then, taking hold of the mandible, they wrench the two sections apart.

Clotted blood is collected in a gourd to be used later for magical purposes. The men then clean the insides of the pigs with bundles of ferns and hand the intestines to our children to be washed in the stream.

I remember in my childhood sitting with my friends on fallen trees in the river, chattering like birds, laughing and playing at our task of washing the intestines until they were clean, and returning full of expectancy to the feast.

The last part of the age-old method of butchering now takes place: that of breaking the pig's pelvis and removing the ribs in one piece. The rest of the meat is cut up into long strips and hung on poles specially erected for this purpose, alongside the fences where the children are hanging curtains of cleaned entrails.

The great fire is now so hot that some of the stones pop and crack, shooting fragments into the air. We scatter with noisy shouts of fear and excitement, but not before one of the men is hit by a burning fragment. Rushing in with ironwood tongs two metres and longer, the men quickly remove the hot stones to the steam-pit, piling them in a mound in the centre of the grass.

We hurry forward to throw wet branches and leaves on top, layering them between the stones. Then, arms full of edible vines, ferns and sweet potatoes, we keep rushing in and out of each other's way, talking and laughing as we set them on the hot stones. Our skirts and *nokens* fly outwards from our bodies as we move in magical unison, dodging the children, dogs and chattering old women who hover too close to the steam-pit and fire.

When all the vegetables are layered on the stones, we circle the mound and gather up the long grass spilling over the ground from the bottom of the pit folding it on top of the pile, trapping the steam inside. After a moment, we unfold the grass to let the men put on the pile the pig meat, skin uppermost, along with fern bundles of entrails, while we wedge in small hot stones.

When the men have finished their task, with much teasing and laughter we gather around to fold the grass back over again, holding it in place while others take long coils of rattan and firmly secure our steaming haystack.

A gentle rain begins to fall, turning the steaming mound into a shimmering green hillock. The steam rises and hides us from each other's view.

Tired, we sit sheltering from the rain, talking while the food cooks. At last, late in the afternoon, we open the pit. The men remove their meat and take it to an area near the chief's hut. Placed on prepared banana leaves, the meat is divided with exact fairness, each man receiving from Pau his share according to his position within the tribe.

We remove our *hibiri* out of the steam bundle, along with the ferns and vines, and divide the food amongst ourselves, putting most of it in our *nokens* to be eaten later.

Sitting near the pit with our children, we hungrily eat what is left, tugging at the vines and sweet potatoes with our strong teeth. The dogs leave us for better fare — the droppings from the men's feast.

All of us eat, ready to burst. When the men finish their food they wipe each other with the pig fat until they glow and their hair shines with myriads of oil beams.

Gradually, with full stomachs, hands and *nokens*, we drift back to our own villages and creep into our warm huts to sleep away the feast day.

The Brine Pool at Iluerainma

3,000 metres above the Baliem Valley, we collect salt as our forefathers did hundreds of years ago. In our own unique way, we follow the pattern of centuries.

Iluerainma, a brine pool in the central highlands of West New Guinea still produces salt. A deep saltwater spring forming a small pool between limestone and sandstone rocks provides the Dani, and other highland tribes with our precious commodity salt, most of which we use in trade.

Climbing from the valley floor in the early morning we travel along a well-worn track which winds through the forest, over slippery rocks to the hanging bridge, where we rest taking our heavy loads of banana stalks from our backs.

Some of us fill our water gourds from the rushing river for the way is long and hard. While we visit, our children play in the water. How I wish I could join them but the path still journeys upwards.

By the time we reach the pool the sun is high in the sky and we are hot and tired, but much work must be done before we begin our return journey.

At the pool we shuck off the outer layers of the banana trunks pounding and squeezing them between our hands so the fibres will become quickly soaked in brine. It has been a while since I visited the pool; the liquid tastes so good as I scoop it into my mouth.

Sometimes, if the work of pounding the fibres is too difficult, we'll ask one of the men at the pool to help. Then we women put the pulp on top of the floating outer layers. We also gather bunches of forest ferns which we tie into bundles and leave bobbing on the surface of the pool. Leaves, fibre, ferns and other greens are crushed, immersed and turned continuously.

As the pool is not very large, only six or seven of us can work at one time. Others arriving have to wait their turn. They wait politely on the huge boulders all around us, chatting and rolling homegrown tobacco in leaves (wild spurge) gathered on the journey.

Turning away from the pool, I wander around the small amphitheatre. Behind the brine pool runs a narrow stream of sweet water. How good the water tastes after our hard work, particularly if there are no men to pound the banana trunks! Beyond the stream partially hidden in the ferns is a sandstone rock. Deep grooves are etched into its surface where the ancient ones had stopped to sharpen their stone adzes over the past centuries.

While I have been away the banana strips soaking up the salt have turned a purple hue. We gather up the fibres, squeezing out the liquid, and then immerse the fibres again. One of our older women is scooping handfuls of the saltish liquid into her mouth, along with pieces of the banana pulp, her mutilated fingers darting from water to mouth like the dancing dragonflies.

A gentle rain starts to fall. Bending over, our brown arms rain-splashed, we hurriedly gather the soaked fibres, all the while stirring the water so that the rain will not dilute the brine. We loop each bundle of fibre around a stick until it can hold no more and hoist the stick onto our shoulders, while young girls fill the drinking gourds with brine. As we leave the men from the highlands take our place.

They are easily distinguishable from our Dani men by the rattan hoops circling their bodies. Only the men come from the high villages to collect salt. The journey across the mountains is not easy for their women; nor is it allowed. The men make the rules, so it is fitting they do the work!

Journeying from the salt pool down to our valley is not easy as the rain is now lashing down. Mud and jungle debris rush over the worn, slippery rocks; waterfalls roar jubilantly on their downhill descent.

At the bottom of our climb we meet up with some of our men, who join us in sitting out the storm under the shelter of the market roof at Jiwika.

After the rain stops and the sun comes out again, we will hang the banana fibres over our village stockade to dry. When the sun has drunk up all the moisture, we will put the dry fibres on a banana leaf on the ground and set it alight. The ash which is left behind will be mixed with water poured from a hollow-gourd to make a dark grey paste. Working the paste well with our hands, we squeeze it into balls weighing about a kilo each.

Once the ash has been formed into balls we weave a cage of dried forest twine around each ball of salt leaving a long piece of the twine, which is knotted into a loop for hanging the precious salt in the kitchen hut to be used by the villagers.

Outsiders may say that our way of collecting and making salt is hard, but it is the only way we know, and the way of our ancestors who for centuries have traded the salt balls amongst the lowland and highland tribes for other commodities such as stones, feathers, pigs and adzes.

The Dani Myth of Bird and Snake

Snake and Bird had an argument about death. Snake held that people die, are cremated and then return to life.

Bird argued "No, that is not good, they should stay dead, and I will smear myself with mud and mourn in sympathy". So men die and birds mourn.

To this day the Dani people believe in the myth.

A Dani Funeral

As he sniffed and wiped both mucus and tears from his sorrow-lined face he leaned towards me. "She was the first of my four wives."

The lean brown body shook with sobs. "When did she die?" I asked. "Yesterday when the shadows were long," Dureg answered. After what seemed an endless pause, he continued. "For half a moon she lay sick. The women of the tribe danced to chase the ghosts away, but she died of bleeding." His voice now but a whisper close to my ear.

"Today I must go and give the news to the other villages so they can attend Mulegai's funeral feast at sun up. The ghosts must be appeased and her spirit freed before the day ends. You and Wynoco must return tomorrow after the moon has left the sky." This meant early morning.

Raising himself from the ground Dureg touched both of our hands, ambled across the courtyard and climbed the entrance stile to the village and disappeared from our view.

My companion Wynoco and I had been walking along the river-bank of the Baliem River where on visiting with some Dani friends we learnt of the death of Mulegai Chief Dureg's first wife from the village of Senamtma.

Since the funeral was to be held the next day we had hurried on our way to pay our respect to Chief Dureg.

Very early the next morning Wynoco and some of the members from Soroba and I left for the village of Senamtma. On the way we passed two women in a wayside pool cloaking themselves in yellow clay in preparation for the funeral. Bedecking themselves in clay is the Dani way of showing mourning. The clay is not removed but it left on the body to wear off naturally.

When we came closer to the village we joined the procession of mourners. Some of the male relatives and friends carried large decorated *je* stones on their shoulders. Others shouldered trussed squealing pigs, as well as traditional *nokens*, and *jerak* (woven bands decorated with cowrie shells), to be given or exchanged at the rites of the dead.

The entrance to the village had been boarded up so we had to climb over the grass-topped fence into the courtyard where Dureg stood awaiting us. As we entered and walked towards him he stepped forward and embraced each of us. We were the first to enter the courtyard since his wife's death.

The women, looking like Halloween ghosts in their mourning mud, sat separated from the men at the entrance to the cooking hut. Inside the hut was packed with close relatives of Mulagai sharing their grief.

After Dureg's greeting the men placed their gifts in a circle in front of the chief's hut and stood quietly, heads bowed. Not a sound was heard, child and dog alike seemed frozen in time.

Breaking the silence Dureg called out, "Mulagai, Mulagai, my first wife, I thank you for caring for my children. You will not be forgotten, many villagers will remember you for you were kind and helped everyone."

As Dureg spoke the Dani language Wynoco whispered in my ear the Indonesian translation, though it was filtered through the sobbing and sniffing of the group. While he spoke the men standing now, rubbed their thighs with clenched fists and shuffled their feet in their grieving.

Dureg continued to call out the good deeds of his late wife, how many children she had had, pigs she had cared for and the general gossip of her life. All the time the low dirge of the mourners created a hypnotic atmosphere.

Suddenly his speech ended and with it all the sobbing. Everyone sat down on the grass-strewn courtyard and began visiting, laughing and similing with each other. When the next arrivals came the speech and mourning began again. In the flick of an eye sorrow-lined faces were creased into smiles.

By the time the courtyard and cooking hut were filled with visitors it was past the time of no shadow. Unable to get into the cooking hut many women sat outside with their children while their *nokens* full of sweet potatoes, bundles of succulent leaves and ferns were propped in tangled array at the entrance to the doorway.

The steaming pit had been cleaned and layered with the fresh sedge grass. Formidable piles of wood lay balanced on the large cooking stones in readiness to be lit. These would be used, when hot, for the steaming pit.

Dureg stepped forward and raised his right arm in the air giving the signal for the pigs to be caught and killed. The village's peaceful calm was shattered with screams of terrified pigs as they careened across the courtyard trying to evade their captors. Larger pigs lying trussed on poles joined in the general raucousness.

The commotion of caught pigs, screams from those wounded and the dying grunts of others made it impossible for me to hear Wynoco's whispered explanations. My eye noticed the deadly aim of Dureg's arrow as he felled one pig after another. Noticed too the children playing amongst the upheaval of the village, and the quiet exchanges between the elders.

As each pig died it was laid in a row facing the cooking hut with fore-paws overlapping each other. Ears and tails were removed and sent to the men's hut to be used later in sacred rites.

The firestick was rekindled and when the grass glowed it was carried to the wood pile. Fresh ferns and banana fronds were strewn on the courtyard floor.

As the flames leapt from the wood the men carried the pig's carcasses over to singe off the hair from the skin which was then scraped clean. Dureg then stepped forward to supervise the butchering of the pigs.

While this was going on I wandered over to the cooking hut and crept inside. Behind the cooking hearth Mulagai's body was strapped into an improvised chair made from banana stalks. Her jaws had been bound with natural fibres, while from her head her carrying-nets hung down on her chest. She appeared so small and fragile in death.

We called out to her spirit. "Mulagai, Mulagai, why did you leave us? We need you, your children need you, too." We wept and hugged each other letting the sorrow flow between us.

Outside the preparation of food had reached the steaming pit stage. People moved quickly carrying hot stones between ironwood pincers. Others piled on wet leaves in the steaming pit as the stones were pushed into place.

Children dogged the milling crowd, dogs sat hoping for a morsel from the feast to fall their way as the men carried strips of pig meat to the pit for cooking.

At last the dervish movements of the throng came to a halt as the long grass draped on the ground was piled up on top of the bundle of sweet potatoes and meat. The bundle was then secured with long ropes of rattan and left to cook and a hush descended over all.

At the opening of the pit a general hubbub broke out. Men carried the steaming meat to the fresh ferns on the ground and began dividing the meat using razor-sharp bamboo knives. This was then placed in piles for distribution to the young boys and men. The former sat like statues in a circle with their dogs in the middle patiently waiting for their piece of meat.

It is taboo for Dani women to eat pig meat unless it is a marriage feast so they distributed the sweet potatoes amongst themselves giving Wynoco and me some to add to our pieces of meat. I was given the honoured prize of pig's liver. I fumbled with the honour bestowed upon me cutting off a small piece with my penknife to eat and while no one was watching, slipped the rest behind me for any stray dog to remove the evidence.

Eating was carried out in silence except for the slurping sounds made by the diners. After the feast the men rubbed themselves with their greasy hands until their bodies shone like brass.

The time had come for the important male guests to join Dureg in a circle for the decision of gift giving and exchanging rites. Except for pigs and *nokens*, all other gifts brought to the funeral are exchanged. This transaction of wealth is witnessed by all present and follows the same pattern commemorating funerals or weddings to re-emphasise social ties within the tribe.

With the distribution of gifts over everyone crowded to the centre of the courtyard where a large square of wood marked the pyre for Mulagai's corpse.

While the flames rose into the air the women in the cooking hut had greased the body. To the sound of durging and sobbing Mulagai's body was carried from the hut and laid on the freshly cut banana fronds steaming on top of the pyre.

Dureg moved closer to the pyre calling, "Mulagai, Mulagai," words lost in sobs as he covered her body with a banana frond. As soon as he did that two other men moved to the burning corpse, one holding a bundle of grass high over his head in order to catch the ghost as it left the body. His companion then grabbed the bundle from him shouting to the ghost to follow him as he ran through the courtyard and out of the open entrance.

As a light rain fell many of the guests began to drift away to their villages. Male relatives of Dureg and his late wife crowded into the chief's hut to talk and smoke until the wee hours while the women retired to the cooking hut to visit.

As I left the village I took a last look at Dureg. There he stood, a grief-stricken man, his brown body wet with the heavy evening mist framed against his warm round hut.

In the early morning hours one of Dureg's wives crept from her hut. Going over to the ashes of the pyre she removed Mulagai's bones and carefully placed them on a banana frond, folding them into a small package. This she carried to the back of the men's hut.

Placing it on the ground she went to Mulagai's hut and withdrew her drinking gourd. Returning to the bundle of bones she drank from the gourd and spat the water over the bundle. This done she hung the gourd upside down on a stick placed at the head of the bundle. With this final act the funeral rites and appeasement of Mulagai's ghost had cleansed the village of Senamtna.

Today, Dani funeral rites in the villages follow this traditional way. However the exchange of wealth has vastly diminished in magnitude, along with body mutilation, and the extended duration of time given to these funeral rites. Animistic and Christian beliefs have become melded together as witnessed following a funeral where the bundle of bones was enshrined in a replica of a Dani hut whose door bore the sign of a cross giving credence to the Dani's acceptance of a new era.

Dureg stands with friends and family members mourning the loss of his wife Mulagi.

Je stones used in exchange rites at funerals lie on top of their sacred nokens (woven fibre nets).

Bedecked in traditional finery of feathered headdresses, massa snail shell collars, pig's tusk and fibre bands, Dani clansmen relax during ceremonial duties.

Opposite top: Highlighted by the sun, irrigation canals of the Dani sweet potato gardens cast geometric patterns upon the Baliem Valley floor.

Opposite bottom: Using her digging stick to remove the soil from the sweet potato's cage-like roots, a Dani woman harvests the evening meal.

Sweet potato gardens are often found high on the mountain slopes above the Baliem Valley. Unlike the Dani gardens of the Valley floor, where river water is abundant, the mountainside gardens depend on rain and the cooler temperatures for moisture for the growing plants.

Covered in clay symbolic of mourning, a Dani woman, burdened both with sorrow and sweet potatoes, trudges homeward.

Amongst elderly tribeswomen one often sees mutilated fingers. Once tradition required young girls to have the first finger joint amputated with a small stone axe by a male family member. Now, in remote areas, the women themselves may choose to have their finger joints amputated. These severed joints were dropped into a small woven bag, and then hung in the cooking hut until the next day when the bag and contents were placed on the lighted pyre at the funeral ceremony.

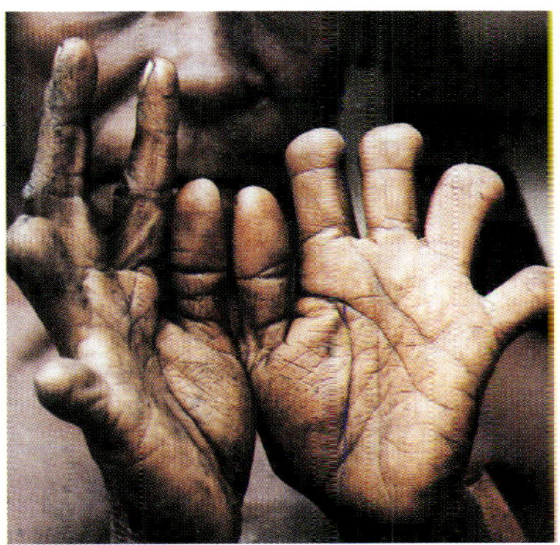

Opposite: Carrying a decorated *je* (basalt stone plinth) used as exchange gifts at marriage and funeral rites, two elderly gentleman stop for a chat on their way to a friend's funeral.

Inside the *hunila* (cooking hut) a bride patiently stands as the orchid fiber braid is wound around her lower body forming the traditional ("yoke") marriage skirt.

On the first day of the making of
the brides' *jogal*, pigs' ears and
tails, usually reserved for the
chief of the village, are placed
on a banana leaf and brought to
the brides' mothers in the
cooking hut.

The mother — her face
whitened in mourning — and
grandmother of a Dani bride
weep over their loss; the young
girl is marrying a man of another
village far away, and they know
they may never see her again.

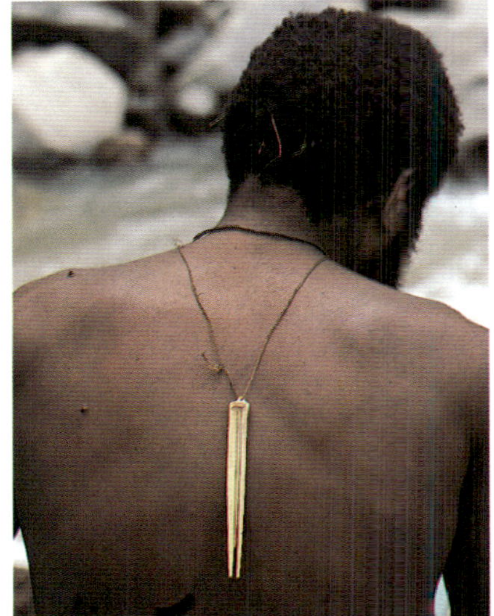

Bigon, a small split-bamboo mouth harp, kept handy for use by the wearer.

An elderly Dani serenades the author outside her hut, with music from his *bigon*.

Opposite: Cockatoo feathers waxed with pig grease adorn the finger-woven headdress of a Dani man.

A mixture of pig grease and soot rubbed into the skin enhances the appearance and creates a fine impression. Pig grease protects the skin from mosquitoes, and is also thought to have magical qualities.

A downward-pointing boar's tusk, worn by a Dani warrior, signals he is on the warpath!

An upturned tusk puts one at ease, as it symbolises peace.

In the past, on reaching manhood, the young men of the tribe would have their septa pierced in order to insert the decorative double boar's tusks.

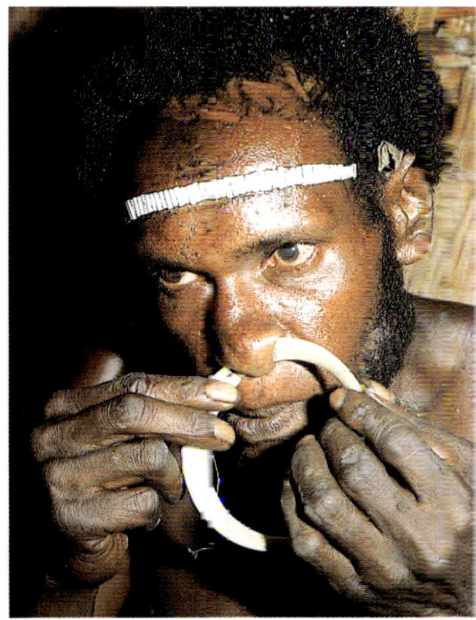

The male populace of the Dani tribe spends many hours decorating their bodies. Headwear in the form of *kuds* or headdress are created out of feathers inserted into either a rattan coil or hand woven hairnet. Even feather dusters are used to creative advantage!

Opposite: Decorated in a warrior's finery Pau poses outside the men's hut in the village of Soroba.

A strange encounter — a tribesman sheltering from a sudden downpour under an enormous rainbow umbrella.

Two friends on their way to a pig feast stop to chat.

This page: After cleaning the bottom of the steaming pit, the men line it with sedge grass. The long strands spread outside the pit are gathered from the ground and pulled over the pit's ingredients. To hold the grass in place, coils of rattan are wound around the outside forming a huge bundle in which the food steams for a couple of hours.

Opposite page:Using ironwood pinchers, hot rocks are placed on top of succulent leaves to produce the steam in which the pit's contents will cook.

 Moving quickly the women of the village set sweet potatoes, root vegetables and edible greens on top of layers of leaves and heated rocks. (refer to page 74)

Opposite page: Stripping off the outer layers of banana stalks and crushing them in their hands, the women soak the fibers in the pool s brine. After the strips become purple and limp, they are removed and placed on nearby rocks to drip dry.Wrapping the salt filled fibers over long sticks, the women brace them on their backs and journey down the mountain to their villages. (refer to pg 75)

78

Inset: Mulikak spiders often become pets of the Dani children.

Opposite: Holding the spine of banana leaf in his hand, Siba sings to the spider vowing not to harm it. As the spider leaves the web, Siba removes all the debris while rotating the strong sticky web onto the spine with his other hand

In order to preserve spider webs so that they can be worn as accessories, the men use the sap of the sweet potato leaves. They crush the leaves to obtain the sap and then rub the spider webs with the sap. The men ease the web from the banana spine with stick tweezers. The strand is then folded over a braided fibre string and fastened with a cowrie shell. Spider web collars are worn in the belief they protect the wearer from marauding spirits.

Carrying a trussed pig between them two men cross the swollen Baliem river via a vine-and-wood-planked suspension bridge.

In this timeless domain of tranquility the stone age people of the Baliem Valley are at one with their majestic surroundings.

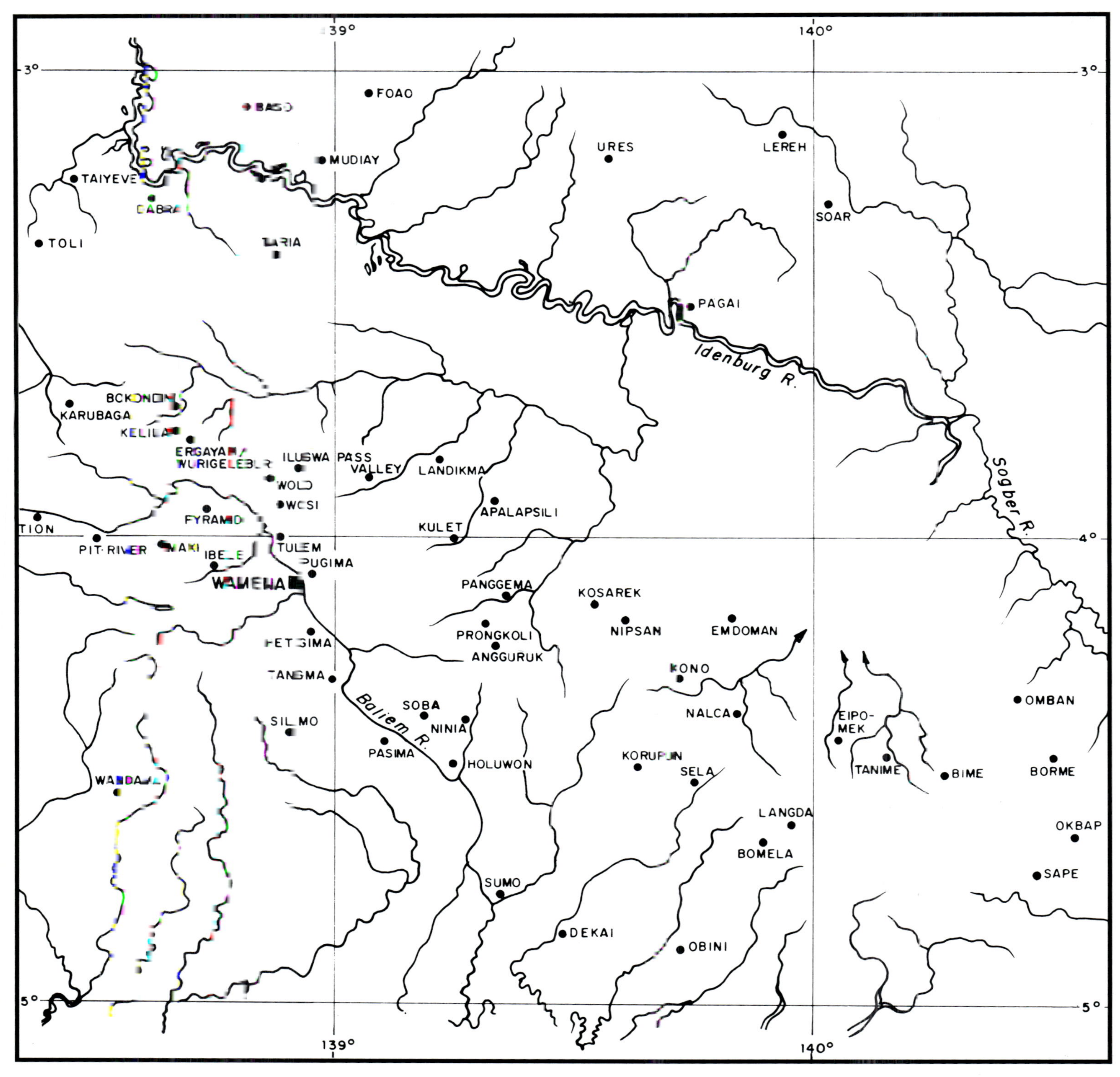

Jale — Central Highlands Region

The Jale of the Central Highlands

Virgin forests, glistening waterfalls and rocky mountain terrain is home to a group of people known as the Jale, who eke out a meagre existence from the hostile upland slopes.

Living in small, conical, pandanus-roofed huts, their world is seldom visited by outsiders. Their daily meetings with teachers from the Summer Institute of Linguistics bring them into contact with a world they will possibly never see.

Local missionaries introduced the Jale to Christianity, and in the small, one-room open-sided school house affording a grandiose view of their mountains, Bahasa Indonesia is taught to the villagers.

Late in 1987 I had the opportunity to spend a few days with the Jale in the highlands of Irian Jaya.

Had I travelled on foot, my way would have led me across rugged mountain ranges and through thick, unfriendly vegetation, across precarious rattan bridges slung between gorges, or over rushing, foaming rivers — a journey I would have very much enjoyed. But time was not on my side, so I took to the skies in an MAF (Mission Aviation Fellowship) Cessna. Leaving the tranquil Baliem Valley, we flew high through the pass and into a world of forest-clad ranges.

Arriving at Kosarek, we were greeted by a band of Jale villagers. Unique in their gyratory rattan hoop skirts, from which red-tipped, long penis shields peaked up and out, they gathered around our plane.

Small of stature, with chocolate-coloured skin and cropped, corkscrew hair, these once war-happy people now spend their days farming, hunting and gathering from the surrounding mountainside.

Great distances and rugged terrain separate the Jale from the Dani in the Baliem Valley. It is rare for a woman to trek across the mountain tops. As a result she knows of little other than her immediate area, which becomes her life. Here she raises pigs and children, and tends the sweet potato gardens.

Jale men-folk go into the forest to hunt small animals and birds, eating the flesh at the kill, or bringing it back to the village for a feast. Every part of the bird or animal is used — feathers and fur as well as teeth and claws become body decorations for the men.

Bare-breasted, the women-folk generally wear only a strand or two of animal teeth around their necks, and a ruffled micro-mini skirt of dried grass. Long carrying-nets, dyed in soft colours from vegetable dyes, hang down their backs, when not filled with a baby, piglet, or sweet potatoes.

Until the age of three, children go naked. After the early years until puberty, young boys wear a single strand of rattan around their waists, while little girls wear the *kam*, a single ruffled, hip-hugging garment open at the front and back.

As the children mature into adulthood, their dress becomes more ornate — at least for the men. Rattan, gathered from the lower forests, is cleaned, dried, and later worn as hoops around the waist. At night this uncomfortable-looking garment is taken off and hung up in the hut, where it resembles a spiral cage. The tribal chiefs are distinguishable by a longer hoop skirt which falls from the armpits to the knees.

The females of the tribe, after the age of three, add a new ruffle to the *kam* every four years, until they become of age to marry at around fourteen.

Sexual abstinence is practised in that after a woman gives birth, her husband will refrain from sexual intercourse until the child is weaned at the age of three or four years. In some areas a man will sleep in the men's hut at least every other night, and when his wife becomes pregnant, he will live only in the men's hut. Today these sexual restrictions are still observed.

The family usually lives under one roof; the men's hut, or sacred house, is used for visits and 'men only' discussions and preparations for ceremonies.

Young boys enter into the world of men after the age of eleven years, when they start to participate in adult male activities such as heavy garden work, guarding the pigs and hunting.

Little girls stay with their mothers, learning to weave carrying-nets, tend the gardens and help with small house-keeping tasks. Most Jale girls many before their first menarche which, in the Irian Jaya highlands, usually occurs between seventeen and nineteen years of age. At this age, a young woman takes on more formidable tasks, such as child and pig raising, as well as helping her husband to plant and weed the fields. The family is the nuclear unit of each hillside village.

By the time the plane had unloaded its cargo and departed, the mountains were hidden from view by storm clouds. We hurried over to the school house for shelter and while we were there, the Jale men took advantage of the storm to make themselves at home around us and our belongings.

Darkness descended suddenly with the deafening noise of rain upon the metal-roofed building. As the noise lessened, one of the men started to play on his *bigon* a small, split and flattened bamboo mouth-harp which is held over the lips while the other hand plucks at a string, causing the outgoing breath to vibrate in the air. Deep baritone voices filled the room with tribal songs and the world of the West disappeared from my being.

After what seemed hours the storm abated and my new friends slipped into the black, star-filled night. My mosquito net had turned into an entomologist's dream: drawn by the light of the lantern, myriads of delicate, colourful, winged and crawling things had taken up residence on the net. In fact, there was barely a spot that was not occupied. Colours of the rainbow shimmered over the white netting, making it difficult to escape into the safe interior to sleep off an exciting evening.

Night gave way to a golden dawn. Rivers and birds vied for supremacy in heralding in a sunlit day. Already, people were returning from the forests with ferns, roots and small birds for the family meal.

My friends of the previous night waited to accompany me to their villages. The going was slippery and difficult after the heavy rains. We crossed small foot bridges made of a single log and slithered over slate-sided waterfalls, dodged pigs, pigs and more pigs who either lay grovelling in mud pools, or grunting and foraging for food in the wet soil.

People passed us, some carrying small pigs. Young girls with blazing smiles had toddlers perched on their shoulders. For the first year of his life, a Jale baby sleeps at night in his mother's arms and is carried by day in the net bag on her back. Until he is about three, he is either carried in this way or on the shoulders of a sibling or parent. After three he begins to walk alone and has less bodily contact with his mother. Everyone we met grinned a welcome; some tagged along until we looked like a miniature expedition. All of them wanted to help carry something — even me, as I slipped from bridge to muddy path.

After a few hours of travelling in this manner, we finally encountered our first village. Small, conical roofs of pandanus leaves and casuarina bark lashed together with dried vines were perched on top of walls of wooden planks hewn from the forest with stone or metal axes. The entrance to the hut was made by removing one of the planks, which, when the occupants were out, was simply replaced. Unfortunately, most of the villagers were away tending their gardens on the far slopes of the mountainside.

Due to the distances and rugged terrain the people build temporary huts in their gardens and stay there while preparing the soil for planting and harvesting the sweet potatoes.

Left behind in this village were some elderly people and a man who was ill. Had I been more observant before entering the village, I would have noticed a bundle of sticks and bark tied to one of the trees to signify that most of the villagers were out and visitors were forbidden entrance to the village. (In the Baliem Valley, a bundle of dried grasses and twigs stuck in the fork of a tree denotes ownership, or that the tree is *wusa*, or sacred, and no one may chop down that tree.)

Leaving the village to the care of its elderly, we journeyed on. The way became very steep. Occasionally we had to descend — in order to climb up twice as high again. Great trees, some moss-laden, blocked our way so that we had to squeeze past, clutching at the lower limbs to stop sliding down the perilous slopes. The scenery was magnificent: below us were low, flat-topped trees; around us, waterfalls, rocks and flowering shrubs whose names neither the natives nor I knew.

After several more hours of toil and sweat we came to a large village of about twenty huts, perched on a slate scree The huts exuded grey wisps of smoke, a welcome sign of habitation.

As we entered, some of the villagers crawled out of their huts to greet us. Small groups of buxom village maidens huddled close together, while children with enormous round black eyes peeked out from behind their mothers' legs. A man wearing a peaked rain cape of dried pandanus leaves bade us enter.

Walking from hut to hut was quite a challenge as the slate was loose and slippery. I stooped down to look inside one of the small huts. In the centre of the floor of dried long grass was a simple hearth on which a few pieces of wood sent smoke curling into the thatched roof. Gaps in the roof and walls allowed sunlight in and smoke out. I was told that mosquitoes were prevalent and that the nights were cold and damp. Some of the luckier villagers had a blanket to protect them, while others simply huddled together, their bodies providing warmth through the cool nights.

In this village, the walls of the huts were not decorated as in the farther village of Anggruk, two or three days away on foot. There, the Jale beautify their huts with stylized, white geometric designs and elegantly arched entrances.

The Jale people belong to the Mek language group, but are less creative than a related tribe, the Kim-Jali, and their lifestyle is based on a less intricate social system.

On the outside of one of the huts hung an old blackened pot together with a length of rattan and a few greens tied together with twine. The daily diet of sweet potatoes and vines is supplemented with ferns and forest greens, poor quality bananas and *siet*, a metre-long red fruit from the pandanus tree. Wild sugar cane is a refreshing thirst quencher and also doubles as a Jale toothbrush, as they rip pieces from the fibrous stalk to slurp the sugary sweetness within.

The women occasionally collect lizards, mice and frogs, leaving the men to hunt with bow, arrow and unfaltering hand, the birds, bats and arboreal or terrestrial mammals found in the high forest Compared to the intake of vegetables, little meat is eaten by the Jale

As with the Dani of the Baliem Valley, the domestic pig plays an important role. It is used in trading, as marriage dowries and for food on special occasions, such as birth, initiation purification, marriage and death.

Near the top of the slope stood a wooden zinc-roofed, single-roomed church overlooking the sea of magnificent mountains of the Jayawijaya range. On our way up we stopped to visit a new mother who had just finished feeding her baby and was making a soft nest of ferns in the bottom of her carrying-net. She settled the child comfortably in his cot, then swung the net over her shoulder, knotted the headband on to her forehead, and covered it with a beaten bark cloth. Picking up her digging stick, she set off out of the village and down the mountain.

On Sundays, after everyone has washed off the weekly grime (mandatory before attending church, I was told), the villagers congregate to take part in a service taken by a Dani missionary. He travels sometimes as far as three or four days between villages to meet the people and conduct the Sunday morning service.

Two colourful events took place during my visit to Kosarek. Very early one morning I saw a crowd assembled outside the house of the teacher at the Summer Institute of Linguistics. Loaded with carrying-nets full of produce for the market, the women chatted unceasingly to each other. Men stood around talking in quieter tones, a few wearing hair decorations and bones through their noses. Some shouldered small fibre stringbags, intricately decorated with small green and red parrot feathers as well as little bobbles of fur Young girls, carrying their younger siblings, joined in the motley throng, while toddlers and dogs vied for a place among the forest of legs.

The local teacher acted as 'financial adviser' — discussing prices with those who had goods for sale, writing on a piece of paper the amount agreed on, and handing it to the vendor, who would then push his way through the crowd to find a space to deposit his wares on the ground. He would set the price tag on top of the pile and settle down to wait for customers.

One man, who had brought a pig's carcass from his village, explained to me as he weighed and priced the meat, that each section belonged to someone, or a family, from his village, and that he was selling their portions for them. Even a small piece of pig fat, used in protective and curative rites, and for greasing the bodies of initiated young males, was priced.

A bundle of forest ferns tied with rattan, a dilapidated cabbage and a long piece of *siet* fruit were priced at about 75 rupiah. The *siet* is cut into pieces and the seeds, which are on the outer side of the fruit, are rubbed together in a wooden trough. Water is added, and the red liquid is then used to flavour and colour the steamed vines of the sweet potato.

Soon the ground was studded with small piles of goods for sale, ranging from vegetables to rattan hoops. I saw some chickens running around, but there seemed none for sale. Apparently the wild birds of prey and numerous village dogs are responsible for the depletion of the chicken population. I was driven to sly but necessary tactics.

One of the Dani men travelling with me on the plane had somehow managed to acquire a chicken. Tied up in a bag ready to he taken down to the Baliem Valley, it looked lost and forlorn. Discussing its plight with the SIL teacher, I decided that Kosarek needed that chicken more than the valley, where fowl are plentiful. While the teacher involved some of the people in a hilarious new game of 'ring-around-the-roses', I slipped the ungrateful chicken to the back of the school house and to freedom. A bleeding finger rewarded my efforts.

Not until a day later did the unsuspecting buyer discover that he had transported an empty basket back to Wamena. I might add that I had seen to it that he would be reimbursed for the loss. After that I seemed to become an all-time rescuer of chickens. On a four-day trek into the far Kim-Jali area I managed to protect our intended evening meal, saving it from its fate each day, until it returned with us still beating its wings.

By nine in the morning, the market was nearing its close; children were getting impatient, dogs fought and a general hubbub took over. The villagers dispersed most of their goods sold. Beautifully woven baskets had been eagerly purchased by the teacher and me.

The highlight of my visit to Kosarek occurred after the market, when the distant thud of drums stirred the breeze. From over the hills a group of awe-inspiring men appeared dressed in ceremonial finery of cockatoo feathers, bird of paradise plumes and high bark hats covered with the skin of the tree kangaroo.

Like gyrating tops, the dancers moved down the hillside. Two drummers wearing long black Bulai bird feathers in their hats, beat upon drums hollowed out of tree logs, painted with geometric designs in red and black. The clicking of penis shields on rattan hoops increased the tempo of the music and set the mood for the shouting, arrow-waving throng of warriors intent on a mock war dance. Their plumes of cassowary feathers decorated with beads and their long nose and ear ornaments transfigured the small, quiet villagers into a wild threatening mob. Small boys wearing hats of wild rhododendrons and what appeared to be red back scratchers decorated with feathers, joined in. The women, not allowed to join the male dancers, rushed around the outside of the group in the opposite direction, their *noken* swaying to and fro as they too helped to turn the morning back ten or twenty years. The scene might have been a raiding party setting out to war.

A small elderly man clad in a simple hat with a single feather, a long white bone through his nose, and a well-worn rattan hoop encircling his wiry body, stood watching with rheumy eyes, the past enacted before him. His young grandson, nearly as tall as him, stood by his side holding a pet chicken in his arms — two generations sharing a lost world while teetering on the brink of a new one.

Opposite: Jale men dressed in traditional clothing of rattan hoops journey from their mountainside village of Prongkoli to visit the weekly market at Anggruk.

Jale women greet passersby as they make their way to Anggruk. The ruffle-like skirts *kan* are worn by females after the age five years. Every fifth year another ruffle is added until she is married, which is usually between fifteen or twenty years of age.

In a remote part of the mountain region of the Jale tribe, families gather to barter for a bride. Placed on the stone cairn are parts of uncooked and cooked wild pig, root vegetables, wild sugarcane, coins and *noken*, all part of the bride price.

Opposite: A young Jale girl clutches one of the family's piglets, as she journeys to the weekly mountain market.

Opposite: Mountain tribesmen are noted for their ingenious bridge-building techniques. Using logs and saplings, hewn wood and rattan vines, they ford raging torrents and deep chasms which otherwise would make their trails impassable.

Below: During high rain season when a more permanent bridge was washed away, our companions gathered small tree saplings and lashed them together with vines securing them to large boulders.Hand rails were then fashioned by arching supple branches to the base of the foot bridge and tying them at intervals with strong vines. All bridges of great span have at each end a sturdy access arch built over steps which lead out to the main bridge structure.

In an environment which is often harsh and unforgiving, the simple beauty of wild flowers has a potent appeal for the tribal people.

Jale women journeying from a mountainside garden rest beside their loaded carrying nets before entering their conical roofed hut seen in the background.

Jale tribesmen carry fire sticks when journeying through the forest. A split stick housing dried grass is bound with a rattan cord then placed on the ground where, positioning it between his toes and using the pulley method he ignites the grass. Decorated with a cowrieshell headdress and bone noseplug a Jale villager coaxes a spark into flame.

Firelight calls for a rendering of a tribal song. Accompanying their *bigon* (mouth harp) player, Jale men produce a melodious whistle by breathing in on a musical note and pushing air out over their teeth - all while pumping their shoulders in an up-and-down motion.

Bedecked in rattan hoops, Jale drummers, wearing cuscus fur hats and feathered noken, beat a rhythm on hollowed-out painted drums. Warriors accompany the villagers in a mock war dance. Females of the tribe are allowed to dance counter clockwise on the outskirts of the manly throng.

Opposite: Once a fiery Jale warrior, this old man, accompanied by his grandson, watches a war dance, which used to form such an important part of village life.

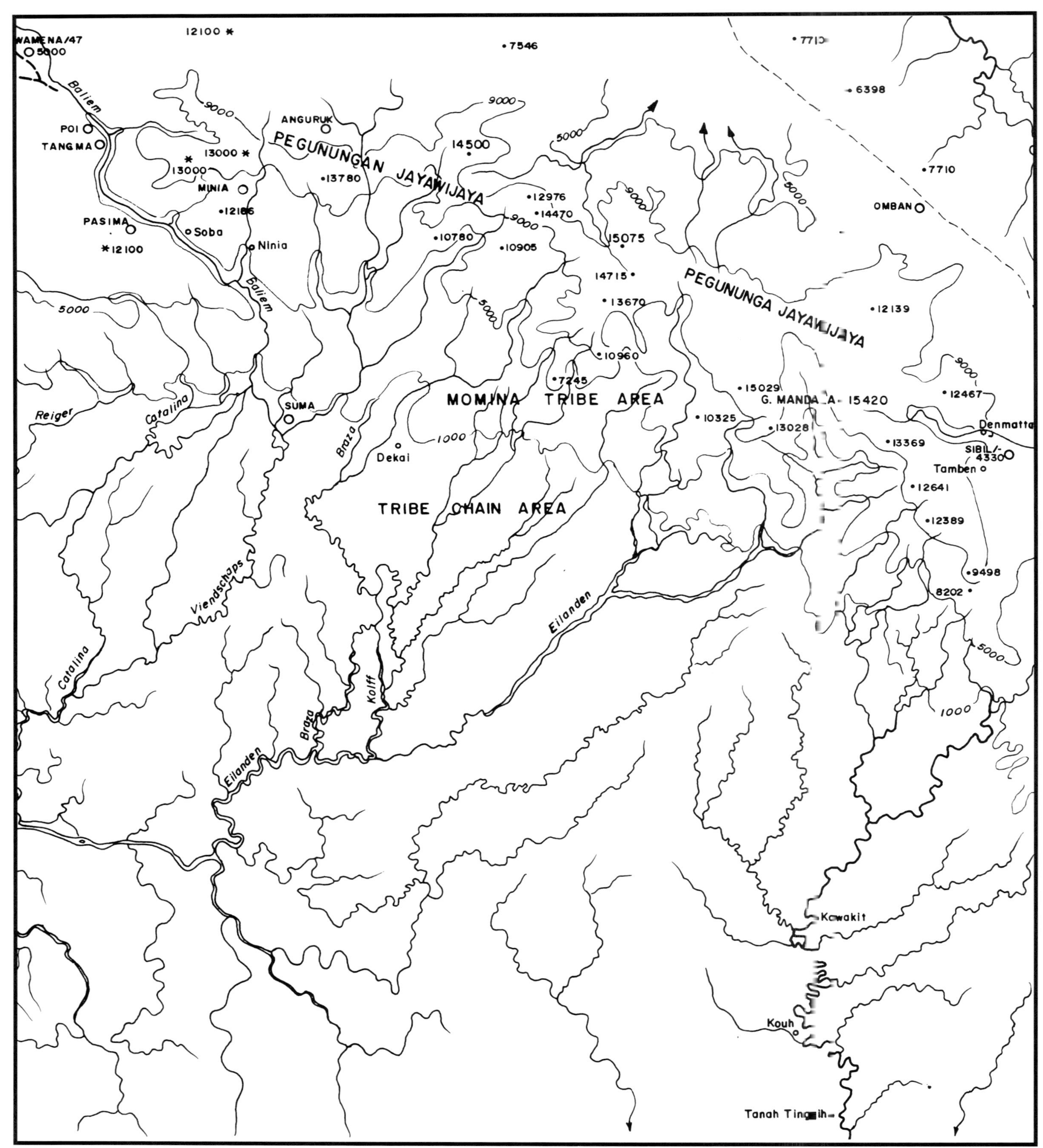

Brazza River Region

Tribes of the Brazza River Region

The Nomads

Rocket-finned tree trunks of unbelievable dimensions dominate the forest. Barely discernible from the ground, their lofty crowns are home to varied species of cockatoos, lories, parrots and other brightly coloured birds. Overriding the raucous early morning chorus, a hornbill drums the moisture-laden air with heavy wings. Bushes and tortuous roots trip the unwary intruder; sago palms and rattan spikes tear his clothes and skin. This hostile domain shelters the nomadic people of the Brazza River region.

The Brazza begins its journey to the upper Sirets River from below the southeastern foothills of the Jayawijaya Mountains. A slow, muddy, winding and switchbacking river, it passes through the grey alluvial plain which houses some of Irian Jaya's most primitive people. To this day some of them wish no contact with the outside world, or even with other tribes.

They build their temporary homes alongside the rivers. Their makeshift gardens lie scattered amongst the felled trees of the rainforest. Simple, elevated structures, their huts are single-roomed, with a small verandah. A single house may shelter as many as 30 people. The houses in some villages are only two to three metres above ground, but further in the interior of the rainforest they are built nine to twelve metres high and are known as tree houses. Access to these, up a notched pole is a hair-raising climb!

A suitably large buttress tree (*Kostermansia*) is selected and felled the desired height from the ground. The house is then constructed on the sturdy trunk base and supported by six to eight large tree poles. The low walls are of split wood and saplings, lashed together with vines.

A slightly sloping roof of pandanus leaves and other rainforest foliage provides a thick covering. Split bark forms the floor of both house and verandah.

Usually the house has one main entrance with a smaller opening for observation purposes. Inside each family builds a river or forest clay firepit on the bark flooring. Often live piglets are tethered to the floors, while carrying nets, bows and arrows rest against the inner walls.

Nomadic dwellers build their temporary village huts lower in height with the front entrance of the men's hut facing the river, where their simple dugout canoes rest within easy reach.

Approaching the village from the rainforest side is a hair-raising experience for the outsider, as one trips and stumbles over felled trees and branches giving the villagers ample warning of one's approach.

In an area close to half a kilometer from the village trees are hacked down to about three meters above the ground, leaving small clearings for the planting of wild bananas, and taro, as well as *sayur lilin*, a broad grass with thin long husks like undernourished ears of corn. Planting is a mere scraping of the ground and setting plants in the meagre forest soil to grow at will amongst the logs.

Since the only approach to the village is by either climbing over the felled trees or walking along their slippery and rotted trunks. It was with great dread I made my journey knowing a missed footing lands one down between the trunks into forest growth with arms twitching in the air begging for Wynoco's assistance.

When a village is abandoned usually after a year or less the cleared areas are soon covered by the encroaching secondary forest growth, covering the felled giants of the rainforest.

Riverside villages consist of up to ten extended family units, as well as the chief's house. Tree house settlements may comprise of just two or three houses in a clearing. Most villages are at least a day's hard walk from each other across wide muddy rivers or deep black streams, through a dank dark world of thick swamp or dense growth beneath towering trees.

The people hunt in small groups, tracking the cassowary, wild pig, tree kangaroo as well as smaller animals. Lizards and insects are either cooked over a fire or squashed between leaves and eaten raw.

Occasionally a giant egg from the jungle fowl is found in the rainforest, its huge nest of forest debris easily discernible.

These big-footed birds build a huge nest about five and a half metres in diameter from the dead leaves and debris of the forest floor. Scratching it together into a mound, the female lays her eggs inside and lets the heat of the mound take care of the hatching of the eggs.

While out in this area one day, I was fascinated to observe some men fishing in the river in a most novel way. They wore home-made goggles with frames of tree bark and lenses of rough glass, glued in place with tree gum and then tied around the head with rattan twine. Diving to the bottom of the river, the men inserted their fingers behind the gills of the fish as they darted out from under rocks or swam in the currents. Sometimes, a small unsuspecting turtle would find itself caught in a bamboo noose by a pair of childish hands whose owner, springing from the river with a look of glee on his young face, would rush away with his prize to his family in the forest village.

Up in the higher area of the Brazza River the people add pandanus nuts as well as wild breadfruit to their restricted diet of sago. Their sago starch, I noticed at a harvesting, was of an orange colour and smelt very sour, unlike the sago gleaned in the Asmat area.

Whenever the chance comes, wild pigs are hunted. A solitary hunter will hide overnight in a tree to await the pig's early morning forage below. A well-aimed arrow finds its mark, the hunter drops down from the tree, hauls the pig onto his shoulders and, grasping its legs around his neck and waist, trots off through the forest to his home.

At his hut his family watch as the hunter sets the carcass down onto a banana leaf, where he is butchering it with a bamboo knife. The precious meat is carefully wrapped in leaves and taken into the hut where a portion is cooked in the firepit. The rest of the meat is shared amongst the villagers or used in trade.

When a sow is killed the piglets are brought to the village where their hind legs are tied with vine to the bark floor of the hunter's hut. When the piglets become too large to stay in the hut, they are removed to a pen underneath until fully grown and ready for eating or trading.

Small and lean in stature, nomadic tribesmen are dark with close cropped curly hair, high foreheads with large eyes set under protruding eyebrows. Their elfin shaped faces are dominated with long splayed noses, full lips breaking open to show small even teeth as a smile like a ray of sunshine flashes across their faces turning the dark eyes into pools of brilliance.

The men's attire is but a few coils of rattan, called *asamko* and worn around the waist. The foreskin of the penis is covered with a piece of banana leaf held in place with a sago palm spine. Ear and nose plugs are either bone, wood or a looped cassowary quill. Rough armbands of twisted bamboo are worn above the elbow.

Some of the men and women shave the sides of the head and part of the crown near the forehead, leaving a tuft of bushy hair which looks like a pompom. Other variations of headdress are long braids of rattan woven into the back of the hair to hang down to the shoulders.

The majority of the women had small holes pierced in the upper portion of the nose in which they had inserted the long thin wing bones from flying foxes. Some of the tree house dwellers of the Kumbi and Korowi tribes had burn tattoos on their upper arms and torso, though this practice is dying out as their lifestyles change.

A mini hip hugging skirt of sago fibres with the thick back panel pulled up between her legs and fastened in the front is worn by a married woman. The unmarried wear a similar garment without the panel. This along with her carrying net and a bone necklace completes a woman's wardrobe.

The men carry undecorated bows and arrows with intricately carved barbed points. Their smoking pipe (*bus*) is made of two lengths of bamboo, one fitted within the other. The much larger base cylinder is covered with ornate geometric designs finely etched with a rat's tooth. In it is tightly packed dry grass, which acts as a filter, and tobacco leaves from their gardens. The narrow upper tube forms the stem of the pipe. Holding the pipe with both hands, the smoker raises it to his lips and draws up the tobacco fumes by sucking on the tube.

The Upper Brazza tribesmen carve the winged roots of the buttress trees into shields, etching basic designs onto the face of the shield which is then painted with traditional white, red and black paint.

Owing to their poor diet and the lack of medical attention, the life span of the nomads is short, averaging, I am told, not more than 35 years. Many of them suffer from *cascado*, an ailment caused by worms that cover the skin in a spiral design, as the fine worms dig beneath the surface the dead upper layers flake off so that the body looks as if it has been dusted in flour. These and other parasites, as well as protein deficiency, all take their toll.

I did not see many young children but of those I did, the babies looked quite chubby whereas the toddlers appeared undernourished. They had probably been weaned at the age of four and were subsisting like the rest on whatever the family could gather or hunt.

The life of a girl child of the Momina tribe is fraught with anguish. From birth until she is five years old she is raised by her parents, living close to her mother and travelling with her wherever she goes. Then at the age of five, she is forcibly removed from her parents and given to another family to be reared by them.

At the age of eight, a man outside the families of both sets of parents may take the child as his wife if he so wishes. He must provide both families with a large portion of pig meat to complete the transaction. Following this, he takes the child into the forest for two or more weeks to consummate the marriage. On returning she becomes part of his household until when in her early teens she is able to bear children and take on the full load of a woman's work.

When a husband takes more than a day's trip into the jungle, his wife must go with him. She is not allowed to remain in the village without him in case she arouses the interest of other men. In the forest she walks a good pace behind him, carrying the bulk of the forest gatherings on her back as well as in her net bag, along with her baby. Any older children often stay behind to be left in care of the other villagers.

Little is known of the lifestyle of the nomadic people. I have recorded here only what I have seen from my journeys into the interior of the Brazza River region and from information supplied by the Irianese people who work and live in the settled villages.

In Search of the Momina

Before the mid-eighties the villagers had little contact with the outside world, until RBMI (Regions Beyond Missions International) sent an

Irianese missionary to Dekai, a village below the Jayawijaya mountains. Now, at least once a week MAF flies to the village, bringing supplies and the occasional traveller like Rudi, my Irianese friend, Wynoco, a long time Dani friend and trusted helper, and myself.

In February 1989 we flew in an MAF Cessna to Dekai in search of the Momina, the most primitive of the nomadic tribes, who live deep in the rainforests. Once across the mountain range east of the Baliem Valley, we flew to the edge of the alluvial plain of Irian Jaya's swamp forests. Soon below us lay a gathering of pandanus tree huts — Dekai.

On our approach the villagers streamed onto the landing strip and when we stepped out of the plane they pushed and pulled in their haste to see us. Swishing grass skirts covered the slender hips yet distended bellies of high-breasted young women. The older women stood apart, their heads bowed low under the weight of their net carrying-bags. Silent and curious, they watched and waited.

Grubby undernourished children and young men in torn and soiled shorts crowded around us, eager to move our things to the one-roomed church cum schoolhouse. Beside it stood a small clinic, manned three times a week for those who might be in need of medical attention.

Everyone that could move into the church with us and watched our preparations for the night. We were not able to get to sleep until well into the night, after finally succeeding in shooing out our audience.

Morning burst its bonds and threw the sun high into a cloudless sky. The teacher lined up eight willing villagers to act as our guides and porters. With their bows and arrows and sweet potatoes, plus our baggage, we filed out of the village to our adventure into the rainforest in search of the Momina.

Fording the first of many rivers, we entered the forest — a barely penetrable world. Above us towered enormous trees, stretching many hundreds of metres in their quest for sunlight; around us thrived a profusion of thorny creepers. Mud pools abounded, as well as great grey swamps. Streams, which criss-crossed our path, had to be forded, fallen trees acting as bridges. Sometimes the trees were very old and broke under our weight, hurtling us into water and mud. One tree-bridge was so high — about nine metres above the river — that the only way I would cross it was by sitting down, straddling the tree trunk and dragging myself across to the other side. None of the bridges

114

had hand holds; with every stream my heart leapt into my mouth and my hands sought Wynoco's reassuring grasp.

Everywhere leeches waved their brown wet bodies and humped obscenely towards us. They worked their way into our shoes and to any exposed parts of our bodies where they gorged themselves until scraped off with a knife whenever we stopped for a rest and a sip of water.

It was 36 degrees Celsius; my camera lens misted over and every view seemed bathed in tears. As the day wore on I paced myself to two words: *dekat* (near) — even if it never was, and *istirahat* (rest) — the five minute break from the gruelling slog.

Finally, coming out of a very difficult part of the rainforest, we came across an apology for a track. Up until then there had been no trail apparent to our eyes. Occasionally we would find a slashed tree sapling and, with heads down on the lookout for the hundreds of roots, follow close at the heels of the native in front of us.

The track through the mud led us into a clearing where a slight figure of a man wearing a loin cloth of bark was working with two women dressed in rolled grass skirts. They were busy harvesting sago. On the ground beside them lay large mounds of orange fibre while in front of them stretched washing troughs of sago fronds. Orange water lay like a cast-off ribbon on the forest floor. Here we rested, watching the three at work.

Setting off again, we reached Kokain, a small ill-kept village of a few dilapidated huts close to a river bank. Fighting the swarms of black sweat bugs which stuck to our skin like burrs, we pitched camp and dived under cover just in time to escape the rain.

Next morning found us on our way again, starting out at about seven to beat the heat and get as far to the next village as possible, not knowing where that was or how far away. We measured distances between villages by using hand signs to ask where the sun would be when we reached our destination. Sometimes it worked; other times not.

If you have never walked in a rainforest, it might be hard to imagine the hundreds of sounds of birds and insects; the dull thud of a tree hitting the ground, returning to enrich the soil from which it had sprung; the pulse of life so strong that you feel a part of it.

When not hiding under leaves, insects swarmed around us. The larger ones, and lizards, I would not point out as I had found out early in my travels that our porters immediately catch them, wrap them in a leaf and munch on the tasty morsel.

That evening we came upon our first tree house. Across a wide river, the house stood against the skyline above the curved river bank. It was a huge dwelling about nine metres high, supported by trees. Here lived more than 30 people. We walked up to the house and rested below it.

Some of the people were returning from the rainforest carrying net bags and young children. All of them had shaven heads with the distinctive pompom of hair on top, and all had pierced ears and noses. Bones from bat wings protruded from the centre of several noses. The women wore grass skirts; the only covering worn by the men was the customary green leaf over the penis, with an occasional man wearing a young hornbill's beak.

The women raced up the precarious single log ladder to the verandah above, where some men were working on their arrows. Other women poked out their heads from the doorway and started shouting and waving their arms at us. Through hand signals, for they spoke no Indonesian, our Dekai porters made us understand that we were not welcomed and must leave at once.

Wearily we climbed down to the river where we set up camp. A storm broke, confining us to our tents. It lasted all night, so we took turns keeping watch on the river which rose closer to us each hour. Finally at daybreak the rains stopped — not too soon as we woke to find ourselves marooned on a small island. The river had risen and encircled us during the night, cutting off any escape had we needed to seek refuge at the tree house.

Packing up was a challenge, not to mention getting off our island. I felt a little uneasy as I pondered the situation. There we were, on an empty stretch of the river, without canoes. No one knew where we were, least of all ourselves. We hadn't heard the reassuring sound of the Cessna since leaving Dekai. We had no knowledge of the trails in or out of the rainforest and were entirely dependent on our eight porters cum guides. Looking back, I realise that we had been more than a little foolhardy.

Later in the morning we managed to cross the receding river and there followed another day fraught with the usual deep mud. However, the varied plant life and enormous butterflies were an exciting distraction. Red and black, yellow and orange, the brilliant blue of the huge swallow tails, all darted around like miniature birds over the streams. We came across insects which we had never seen before, some turquoise, others sporting green stripes. We passed a cassowary nest the size of a small room. The mound of wet forest debris possibly hid eggs the size of a large grapefruit. We came across a bare stretch of grey ooze which, when we tried to cross, sank in over our boot tops. Thin broken spikes of dead trees made the way more harrowing. Occasionally the way ran alongside a river, and for a while my cameras and I would dry out in the sunshine.

After about four hours we came upon another village, a collection of huts on high stilts. There was a hushed eeriness about the place. We knew the village was inhabited, but there were none of the usual sounds of life - no voices, no laughter, no movement. Creeping forward cautiously, we found people in the first house, lying on the floor, their heads covered, moaning softly. Our porters were all for running back to the forest; but we managed to restrain them. Slowly we moved on to a larger hut facing the river. Here sat a stony-faced individual wearing a hoop of bamboo around his waist, a rag on his head and a huge earring made from a cassowary quill through one ear. Through the opening of the hut I could just make out the form of a small child who began to whimper as we approached.

The men from Dekai reluctantly offered the chief the tobacco and salt we had brought for just such a moment. Not batting an eyelid, he continued to stare at us. Silence. After what seemed an eternity, the chief spoke to the porters, telling them that we must leave at once or the frightened villagers would attack with their bows and arrows. Suddenly, a young man leapt on to the veranda, his aggressive intentions unmistakable. We decided it was time to take the chief's advice.

He led us across the wide river, the current so strong and the pebbles so slippery that we had to make a human chain of hands to get across. On the other side we saw his wife coming out of the forest with her dog, which fled howling back into the trees when it saw us. In the net bag on her back was a tiny baby, however, there was no time to observe more. The chief indicated that we should go quickly and remain on the far side of the river. We departed, still wondering about the cowering family in the first hut. Were they merely afraid of us, or were they mourning some private grief about which we knew nothing?

After marching about two hours in the rainforest, we heard much shouting from the river. One of our porters went to see what was happening and returned to announce, in sign language, that one of the men from the last village wanted to see for himself the white outsider.

My 'brave' companions immediately retreated a fair way from me. Suddenly through the creepers and trees emerged a small, very fierce-looking man, brandishing his bow and arrows and wearing a lot of rattan in his hair. He came straight up to me, and I was glad that I had my old green jungle hat rammed down on my head, hiding my long red hair. (I had been told that red is considered 'of great power', and that some native might covet my hair. Of course, my head would go along with it.)

While I stood rooted to the spot, he walked around me, poking and touching, peering up under my hat into my face. After fearful seconds of eternity, I found my voice and shouted to the others for help. The man jabbered something, smiled and accepted our proffered gifts of salt and tobacco, then ran off to the river, laughing hilariously at the funny sight he had set his eyes on.

Another day passed, another brush with the unknown, another night of rain. Each day brought its trials and tribulations. Soon our supplies ran out. Earlier in the trip our porters had taken a liking to our rice and when asked, we felt we should share it with them, in case they decided to take off in umbrage. Unfortunately, we had not brought enough, so we turned to eating green papaya, roasted grasses, sour sago and the odd tiny river fish.

More than the swamps, I came to dread passages through temporary gardens with their half-felled trees that required balance and care every step of the way. One slip could mean a twisted ankle or broken leg, with no help readily available. Half a kilometre through such terrain could take us a couple of hours.

The beauty and hostility of the rainforest never ceased to amaze me. Rivers and streams, cloudy with minerals, threaded through the dense vegetation. Giant ferns stood gulping up the sunlight at the water's edge. The length of a tree trunk might be covered with strange fruit coloured a vivid orange, red or blue. Higher up hung orchids of soft colours. Great blossoms from the canopy above lay like discarded dresses on the forest floor, while across the petals ants the size of small grasshoppers marched past in search of food. Occasionally a snake would slither out from beneath a log, frightening everyone.

Finally, the day came when, we were told, Dekai was just one or two days' trek ahead. Sight of a temporary hut reassured us that we were on the right track. On nearing a village named Keikai, our porters abandoned our baggage and made off. They had had enough. Tired, without food and infected with rather painful sores, we decided that we needed to get back to Dekai as soon as possible. Wynoco went in search of the men and, promising to raise their pay, managed to persuade five of them to return.

Leaving the village, we struck out once more into the forest. The rain came down in sheets, blinding us as we sank and slithered in the mud. Leeches clung to us, logs slipped from underfoot as we crossed streams, the heat was unendurable, the forest denser. I began to doubt the wisdom of pressing on from Keikai.

Walking across some half-felled trees, I slipped and fell head over heels into a gulley between them. Luckily, Wynoco happened to turn around and noticed my disappearance. Reaching down and grasping my waving fingers, which was about all he could see of me, he hauled me back to the light again.

At last we crossed the gardens of Dekai, beside the airstrip. Six days of exploring a tiny part of the vast rainforest had come to an end. It felt like sixty. Tired, filthy and hungry, we climbed the rickety stairs of the schoolhouse and threw ourselves upon the bark floor

The villagers had grown quite worried about us, thinking we were lost. How close to the truth they had been! We bought a piece of wild pig from them and cooked it over an open fire together with some swollen grass buds and sago. Food had never tasted so good.

Sitting around the fire with the village headman after dinner, we learned that we had passed through a series of primitive settlements with romantic names like Kubu, Botonikama, Kokain, Moroku, Indamabolin and Keikai. Should we return a year or two later, the temporary villages would have been abandoned, along with their names.

After contacting the outside world we learnt that we would have to wait two more days before being flown out to Wamena. Using the schoolhouse as our temporary quarters, we spent the time visiting with the villagers and learning about their way of life.

On our last night a huge storm blew up. Wynoco and Rudi moved their sleeping bags to another corner of the room to shelter from the rain. I wasn't getting wet, so I stayed where I was, listening to the storm raging outside until I finally drifted off to sleep. In the middle of the night I awoke to find my mosquito net in total disarray. The edges had been tucked neatly under my sleeping bag before I got into bed, but now the net was a mess, sprawled out across the floorboards. Rudi and Wynoco, their nets undisturbed, were snoring loudly, so I returned to my sleeping bag, tucked the net back into place and went to sleep.

Some time later I awoke with a start. The mosquito net was being jerked from under me. I heard soft whispers in an unknown tongue close to my ears and felt little fingers touching me, moving up my arms. I tried to cry out, but my voice stuck in my throat. Finally I groped for the flashlight and switched it on. There was *no one there*, but the net was pulled out from under the bag again.

Hurriedly, I got up and flashed the torch all around the room. Nothing. I ran to the door and searched outside. No one. The storm had passed and the wind had dropped. There was absolute silence. I called to the two men but they slept on, their nets still undisturbed. Returning to my corner, I spent the rest of the night sitting up in bed under my net, too scared to sleep. Daylight took a hundred years to come.

When dawn finally broke I woke Rudi and Wynoco and related my night's experience. They appeared remarkably unconcerned. "Why didn't you put a piece of lime under your pillow, or in your pocket?" asked Wynoco.

"Whatever for?"

"To protect you from marauding spirits, of course. Everyone does that in the forest."

120

The villagers confirmed that it was "just the ancestral spirits" visiting me because they were curious about the alien in their village. They had intended no harm.

Often, when staying in a tribal village, I feel very close to the world of spirits, but never *that* close. Others to whom I have told this story have said that they, too, have had similar experiences when out in remote villages. In fact, they do not venture out alone after dark.

As I waited for the flight back to Wamena and to the modern world, memories of our arduous trek rippled across my mind, and I felt a great surge of achievement, and pleasure, that I had been born with a sense of adventure.

Resting on the platform of their raised hut, a Momina woman and child watch the hunter butcher his kill with a bamboo knife and ironwood stick used for prying open the jaws. Whitening of the woman's skin is due to cascado - an ailment caused by minute worms burrowing under the skin which then dies and flakes off.

A young Brazza girl playfully peeks out of an opening in the sago leaf siding of her hut in the village of Senggo.

Momina childhood is short-lived since young females are often wedded at the age of eight years old to elders of the tribe. After consummation of the marriage in the rainforest, the young bride lives in her elder husband's hut.

Inside their hut of padani ribs and sago frond thatching, a family gathers around the central firepit for a noon-time meal. Amongst their possessions on the bark floor are beaten bark shoulder capes, firewood and piglets, offspring of hunted sow.

Opposite: Sago starch is brought home from the rainforest wrapped in sago fronds and kept inside the family's hut. As needed, a lump of starch is broken off and rubbed between the hands into fine crumbs, which are piled, onto a banana leaf. Wild fungi, insects, small lizards or pieces of fish are placed on top and the banana leaf folded into a package and then baked in the open firepit.

From the tribe of tree-house dwellers, a young woman shelters from the rain deep in the lowland rainforest of the Upper Brazza River.

Opposite: Waiting for the men to finish digging out the sago pith from the palm, a young mother adjusts the finger woven head net in which her three month old child sleeps on a bed of leaves. In the lower net is a sleeping pad made from dried pandani fronds.

128

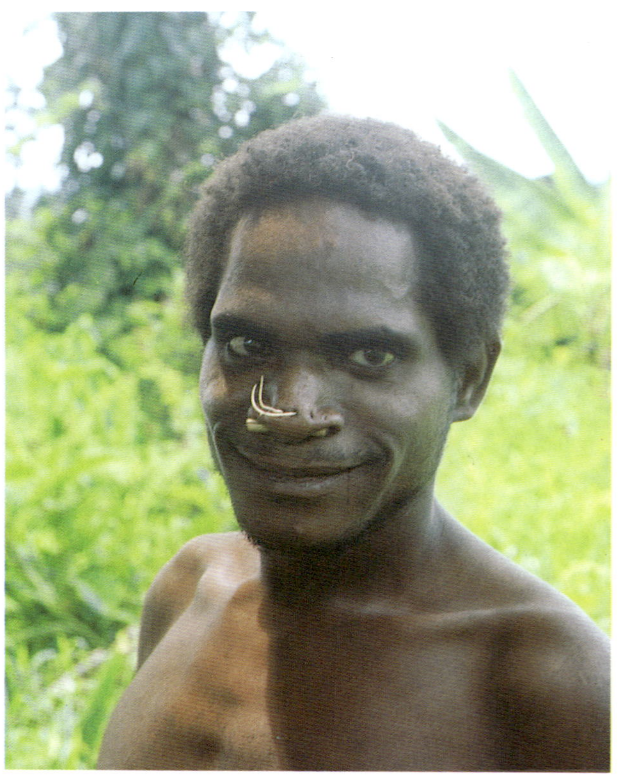

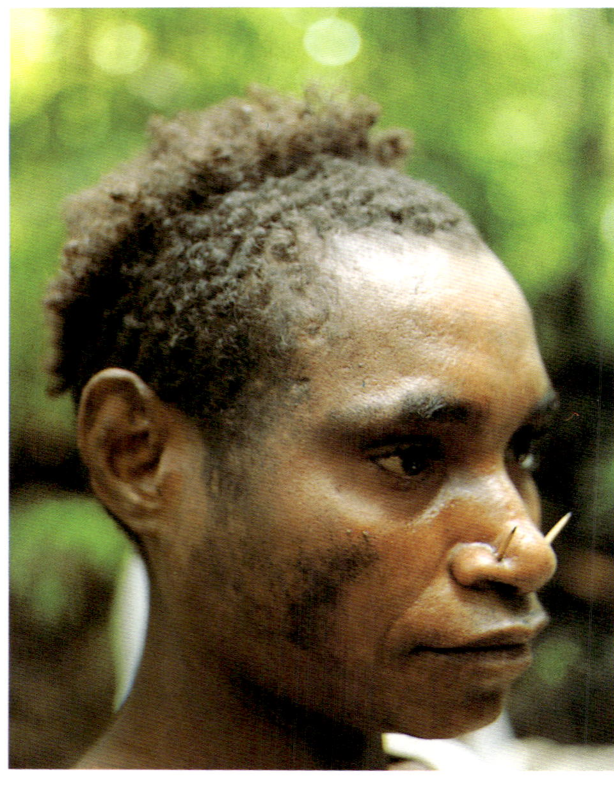

Tribesmen from the rainforest of the Upper Brazza area meet for the first time outsiders from another tribe and a Caucasian woman. Nomadic tree house dwellers pierce the upper part of their noses and insert either bat wing bones or splinters of pig tusks. Their village chief distinguishes himself by wearing long hoops of cassowary quills decorated with broken cowrie shells through his ears. Pom-pom hair styles symbolic of Momina women are cut into shape with bamboo knives.

Opposite: A Kombi woman returns from the night's hunt with the day's food and piglets which she will raise in the communal tree house.

This page: Decorated with bat wing bones in her nose, cowrie shells braided into their hair and wearing a grass skirt of sago fibres, a Korowi woman pauses near the village of Yani Rumah.

Opposite page: Tree houses built up to 40ft high are mostly found in the Becken River region. Home to Kombi and Korowi tribes these shelters made from buttress trees, saplings and woven and thatched with dried pandani fronds, house up to four families.

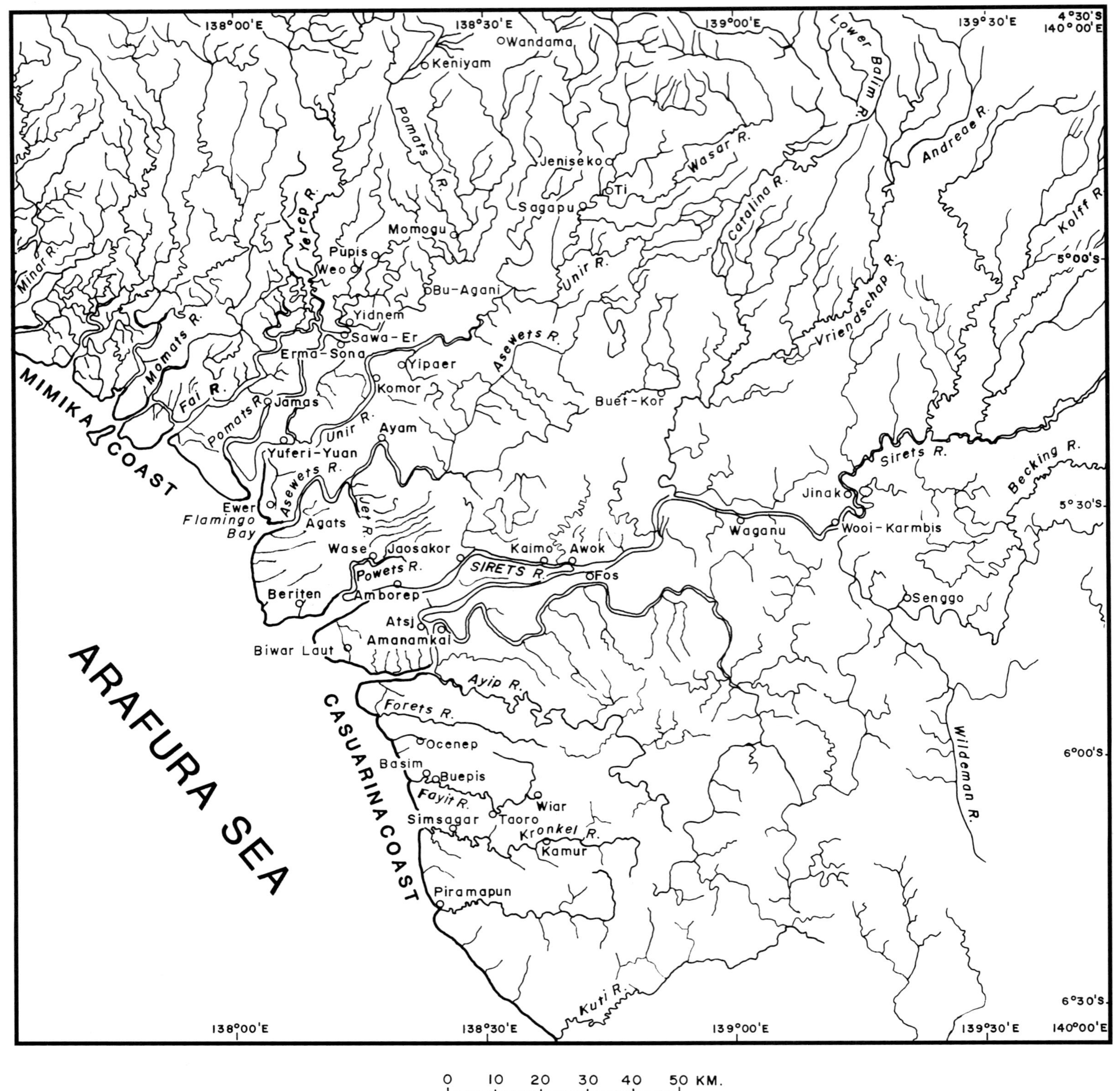

Asmat and Upper Sirets Region

The Asmat Woodcarvers of the Swamps

Below the majestic Jayawijaya Mountains lies the hostile and secret domain of the Asmat people. The alluvial swamps, with their tidal rivers and sluggish black streams, are covered by a canopy of dense rainforest. The coastline is fringed with thick mangrove trees, their roots, like octopus tentacles, exposed when the low tide bares the mud for as much as a mile out to sea.

Life is abundant in this region, but both humans and wildlife are hidden from view by the dense forest. Brilliant birds, cassowary, marsupials, rats and flying foxes abound. The wild boar claims sovereignty over the smaller animals, as the area is not congenial to larger creatures. The rivers are home to the fish, turtle, shellfish and, of course, the lurking crocodile. Whales and sharks patrol the treacherous currents of the Arafura Sea.

In this region live the Asmat, a people who believe that they were created from wood. Decades ago the Asmat people lived in large villages as a protective measure against the perpetual wars among the tribes. Most villages were built in the curve of a river facing the water so that they were not open to surprise attacks. They look much the same today, but the threat of tribal wars has long ceased.

Because the rivers are tidal the houses are built on stilts about one and a half metres high. They are accessible either by canoe or rickety walkways. Huts built away from the water are reached by climbing up a notched tree trunk.

Nipah palms, split sago fronds, and leaves and roots of the pandanus tree form the building materials for the Asmat shelter. All are securely

lashed with rattan strips, including the heavily thatched roof. The floors are of split bark.

Within a communal or extended-family house, each family has its own entrance and its own fireplace of river mud. No furniture is used, so the few family belongings are hung on poles above the fireplace.

Means of Travel

Land travel is hardly known, except for trips into the tangled interior to look for sago palm from which the starch is harvested. This forms the staple diet of the Asmat people.

All other travel is on water. The dugout canoe is the Asmat chariot, considered even more important than the house. The canoes vary in size, the longest being war and family canoes which can measure up to 20 metres. These are paddled from a standing position, the men arranged according to rank. The women's canoe is smaller and more easily manoeuvred, with the women sitting or kneeling to paddle. Even youngsters have their own canoes, measuring only about 2 metres, which they paddle with great dexterity while out fishing or simply playing chase or mock wars on the myriad waterways.

The village chief decides when a new canoe is to be built and presides at the ceremonial feast held in honour of the spirits of the canoe.

The *tsji*, an Indonesian name for the tree used for canoe building, is felled in the forest and the canoe roughly shaped on site. The shaped trunk is then floated back to the village where work on it is continued.

Prowheads are carved by the village specialist, usually a *wow ipit*. The heads represent dead ancestors or animals and birds. When the prowhead is completed, the canoe is covered in a protective whitewash of lime made from baked and crushed clamshells mixed with water. Burnt yellow ochre and charcoal soot is used to etch the carved figure. Vertical stripes are painted in red on each side of the canoe, and tufts of sago fronds inserted in small holes lining the rim.

Under the canoe smoulders a fire of sago fronds to dry and harden the wood and to make it more buoyant. After this process the canoe is inverted. Holding a clamshell, one of the men runs from stern to prow, drawing a line along the charcoaled keel to free the spirit of the canoe. After the residue of the charcoal is cleaned off, the canoe is righted and set in the river.

To celebrate the completion of the work, boat races are held amidst much shouting, whooping and dousing of each other with white clamshell powder. In the general confusion there are no winners and no losers. The day ends with dancing and feasting.

Sago

Sago plays a key role in Asmat life. It is not only the staple diet but a commodity used for trade and purchasing a bride. The wealth of a village as well as an individual is measured by the amount of sago-producing land that village or man holds. Patrilineal ownership controls the passing of sago rights from father to son. Where there is no son, the son-in-law inherits the sago grounds of his father-in-law.

Many tribal wars were fought in the past over sago rights. Today, with the pressures of forestry, some of the grounds are threatened.

To the Asmat the sago palm is indeed the 'tree of life'. Without it, he has no food, shelter or (at one time) clothing.

Dark green and stately, the sago palm grows wild in the lowlands and swamps. The nipah, which grows along river banks, is similar in shape and colour and is often mistaken for a young sago palm.

It takes about 15 years for a sago palm to reach maturity, at which time it bursts into a single crown flower. Harvesting must be completed before the palm reaches this stage, or else the pith is too old and dry for use.

Sago gathering is a village affair, with the children — when not at school — accompanying their parents in the ritual. Leaving the village in the early morning light, the people travel by canoe to a side-stream. There they build a bivouac of nipah palm, where they will stay until the harvesting is over.

A traditional ironwood stick is used to pry open the bark and dig out the pith, and a small fibre screen is used as a sieve. Whatever else is needed (containers, scoops, etc) is made at the site.

The tree is felled about one and a half metres above ground level and its huge thorns are lopped off. While the men prise off the bark to get at the pith, which they will later pound, the women make the fronds into collecting and washing troughs. First they must crack the tough frond spine to make it more pliable. If there is no stream nearby, the women have to dig a well.

Uncracked fronds are pushed into the ground as supports for a series of two or more troughs, erected on a slope. In the upper-most trough water from a clam shell is poured over the pounded pith to sluice out the starch. The liquid runs through a fibre screen into the second trough, where sticks are placed to slow down the flow of the starch-laden water so that the heavier sago starch sinks to the bottom.

The women rush backwards and forwards between the men and the troughs, laden with the pounded pith which is washed and then squeezed dry. The starchy extract from the pith settles at the bottom of the second trough and looks to all appearances like plaster of Paris. It tastes somewhat the same, too! The sago paste is mixed with fruit, or delectable morsels such as lizards, ants and other insects, and the whole 'sandwich' is then roasted.

As much as 35 kilos of starch can be gathered from a single palm. This is collected in funnel-shaped baskets made from sago palm strands. When full, each basket weighs about five kilos. These are loaded into the canoes for transportation back to the village.

Sometimes young palms are felled for the heart, considered a great delicacy; or older palms are felled and the pith cut into strips for chewing like sugar cane. Other palms are deliberately slit to enable the capricorn beetle to lay its eggs in them. Within 30 to 40 days the eggs hatch into plump sago grubs. The tree is then felled with great ceremony and carried back to the village, where the grubs are collected in a special tray painted red and white and decorated with sago palm tufts. They are eaten either raw or roasted over a fire until juicy and crunchy. The taste is not unlike pieces of chicken fat.

Certain taboos must be observed at the time of sago harvesting. Going to the grounds, one must not be asked where one is going, or pass a woman bailing water from a canoe. The axe must not strike against stone. No fires should be made near the palms; dogs and young children must also be kept away. Harvesters must not borrow anything from each other en route to the grounds, and no pork, fish, cassowary or bananas may be eaten.

Customs and Beliefs

The *Jeu*

The *jeu* (men's house) is a rectangular building, sometimes over 60 metres in length. In times past it provided living quarters for young boys, bachelors and elderly men. Here the men once planned their wars and headhunting excursions and performed their sacred rituals. Leaders of the *jeu* were chosen for their prowess as warriors and headhunters.

Today the *jeu* is still a male preserve for kinsmen descended from a common ancestor. They gather on the bark verandah to carve, pass the time of day, and plan various ceremonies of a more gentle nature.

An Asmat sketchbook compiled by Bishop Sowada of Agats tells the following story about a special *jeu*:

Long, long ago there lived a man named Sok who spent much of his time in the village *jeu*. One day he was in the *jeu* singing and dancing in celebration of the Feast of the Mask Spirits when his wife interrupted him. Breaking the strict taboo, she entered the *jeu*. "While you dance and sing, your children go hungry and the fish multiply in the river," she complained.

"Not so, wife, not so. You are mistaken; you are wrong. While the men drum and sing, I must dance. We do not play; we observe the Feast of the Masks. And your coming here brings danger to us all."

But his wife insisted and so Sok left the Feast, went home for his fish trap and husband and wife set off for the river. As he was laying the trap in the water a spirit mask appeared from its depths and bore him away towards the mouth of the big river. Weeping and wailing in terror his wife rushed back to the village.

"What's wrong? Where is your husband?" asked the villagers, crowding around her.

"He has been dragged away by a spirit mask to the mouth of the great River Jii," she cried.

In consternation Sok's family and friends rushed into the jungle to cut wood for a huge trap, which they threw into the river, hoping to save him. But they failed.

One day one of Sok's sons was down at the river when he saw a large *jeu* on the riverbed, and there, in front of the underwater *jeu*, he beheld his father. "Come back, come home," he pleaded, but Sok shook his head.

"I must stay in this spirit house," replied Sok. And not even the pleas of three other sons could move him.

Two lessons are learnt from this: no woman may enter the *jeu*, and everyone must respect the Feast of the Mask Spirits.

Origin of the Tribe

Another legend concerns the origin of the Asmat and explains why to them, wood is the symbol of life:

At the beginning of time there were two young men named Famiripits and Mbouirepits who lived by the Sirets River and were very close friends. One day when out fishing they spied two lovely maidens collecting shellfish on the opposite bank. The women called out to them to cross over, which they did, drawn irresistibly by their beauty.

Eventually each man took a maiden as his lover. However, Mbouirepits discovered that his friend was betraying him with his mistress. He departed in anger to the other side of the river, taking the canoe with him and leaving his friend stranded.

Famiripits mourned the loss of his friend and so great was his grief that his mistress took pity on him and agreed to help him cross the river in search of Mbouirepits. To preserve her reputation, she wrapped him in nipah palms securely tied with rattan. Thus hidden from all eyes in her canoe. As they were crossing the river huge waves dashed against the canoe and overturned it. The woman tried to save her lover but the current tore him from her grasp. She could only watch in horror as the bundle of palms was borne farther and farther away.

142

Struggle as he might, Famiripits was unable to free himself, and so he drowned.

The strong current carried his body down the Sirets and into the Kastel River where it was washed ashore on a small island. The birds on the island pecked open the wrapping and were surprised to find a man wrapped in the leaves. "He must be dead," said the gulls. "Let us eat him."

"No, no! There might still be life in him," argued a kindly hawk.

They finally decided to consult the sea eagle, who coated the body with a secret medicine which restored Famiripits to life. On awakening, his cries of fear frightened all the birds away, except the bold and wise sea eagle, to whom Famiripits told the tale of his adventures.

Realising that he needed shelter, Famiripits set about felling some trees and building the first *jeu* (men's house). Then, needing companionship, he cut down more trees which he carved into statues shaped like himself and his lost love. So lifelike were they that taking his newly-made drum, he beat upon it until the statues came to life.

Unable to settle, Famiripits travelled ever westward, carving new statues and creating new life with his drum. Thus began the Asmat tribes.

Very little is known of the true history of the Asmat, except that they are of Melanesian stock. To this day they identify themselves with the tree, equating the body to the tree trunk, the head to the crown of the tree, arms to branches and feet to roots. Within both man and tree, they believe, beat a stout heart.

Temporal and Spiritual Worlds

In their daily lives the Asmat strive to maintain balance and harmony between their spiritual and ancestral world on the one hand and the temporal world on the other. In times past this was played out through warfare, headhunting and cannibalism, the last in order to absorb and control the spirit of the dead man. Ceremonial feasts are the highlight of every tribe, preceded by hunting and foraging for food in the surrounding jungle.

It is through the skilful carver that the invisible is made visible to the rest of the tribe, in the form of beautifully crafted ancestral poles, shields, masks and canoe prowheads. Each carving is named after a specific ancestor. Not only can the spirit take on material form, but animals can take on human form and vice versa.

Animists, the Asmat believe that *bii* (spirits) dwell in all parts of the environment. They may be good or bad; some are just mischievous, appearing at odd times, scaring and playing tricks on the villagers. Like humans, spirits may be fair or dark, short or long haired.

The *tememer*, a very powerful spirit, has a long pointed nose, eyes that glow in the dark, sharp teeth and long hair and nails. It lives in the ownboro tree and is the spirit of a woman who has died in childbirth. Many other trees also house spirits: *borokomor*, also in the *ownboro* tree, and the *jiaw bii* which lives in iron wood tree hollows, are said to be very dangerous, attacking people with sago sticks.

Bii also live in the rivers and require offerings of food and tobacco — also flashlight batteries in these modern days! Without these offerings, it is believed that the fishing and sago harvests will be poor.

Lonely spirits of the dead often try to snatch the soul of a living family member to take back with them to the other world. When there is a death, a member of the village is paid to wail all day and night in order to frighten away the spirit of the dead so that it cannot capture the spirit of a loved one. The people try to reassure the spirit that it is still remembered by referring often to the good and friendly deeds of the deceased and by wailing in the afternoon and evenings.

If a living person wakes from dreaming of a dead relative, he should immediately start wailing loudly to frighten the spirit away. If it persists in entering the hut, the owner must secure everything and place gifts of tobacco, sago or a small carving outside the door.

No ceremonial wailing takes place, however, if a child dies, as it is believed that the child chose to die in order to protect its earthly family from bad spirits.

Some of the Asmat are well-versed in the art of magic and sorcery. To summon rain, special leaves are rubbed on wooden cooking tongs which are then placed facing outwards under the thatch of a hut. Other leaves with magical properties are rubbed on an arrow point previously dipped in red ochre and lime paint. Then, when held by the shaft, the arrow will point the way to anything of value which has been lost.

Powerful sorcery is passed on from mother to daughter. Some tribal women wear small pouches on their person in which they keep secret items of sorcery. It is believed that a sorceress is able to kill her enemies by shooting them with invisible arrows made from bone (once human bone).

Male sorcery or black magic is centred on a piece of evil-smelling wood. When carved into the shape of a crocodile, rubbed with magical plants and tied to the river bank with pandanus leaves, it will make crocodiles attack and kill any unfortunate person who ventures into, or even near, the river.

An Asmat told me that for some time a crocodile had persistently preyed on the people of his village when they bathed or fished in the river. After much searching they at last found a carved figure of a crocodile tied to the bottom of a pole in the river, and after it had been destroyed no more people were taken by the crocodile.

Feast of the Spirit Masks

On a lighter note, the Feast of the Spirit Masks is a happy event for the village. The masks are named for the dead and a man wearing the mask portrays that particular ancestor and acts out the rcle which the deceased played in life, mimicking even his movements.

In great secrecy in the men's hut, the masks are constructed from woven dried fibres and decorated with shells, beads, seeds and cockatoo feathers. The garment to accompany the mask is a tunic of red and white, adorned with raffia from the hem to the floor and from shoulder to wrist.

On the day of the Mask Feast a 'spirit' makes an appearance from the jungle but is chased back by the villagers. At about four in the afternoon when the shadows lengthen, a villager goes into the jungle and leads out all the masked spirits who sing and dance on their way to the *jeu*, where they are welcomed by the families of the deceased. All join in the celebration, feasting, dancing and singing through the night until the first signs of dawn. As the rays of the sun dispel the darkness, the spirits return to the jungle.

In the upper reaches northwest of Asmat mask dances are still held. On my last visit one had taken place only a few days before my arrival. I was shown a beautifully made mask, and this Song of the Mask was recited for me:

> Oh Spirit, come out from the jungle,
> Follow the one who leads you.
> Dance and rejoice with us all night,
> For you come to bring us strength,
> Courage and fertility.

The Initiation Ceremony

Perhaps one of the most important occasions in the life of an Asmat male is the initiation ceremony which takes place when he is between 13 to 17 years old.

While the women and children of the village remain at home, the men cross the river and stay in a hut where they beat drums and sing through the night. The next morning they uproot a tree and remove all its branches. It is then painted red, white and black, and decorated with weeds from the river. This *ok eyok*, as it is called, is then stood inverted in a canoe, its roots pointing skywards, and paddled back to the village with much shouting and blowing of horns to warn the women so that they can hide from view. The *ok eyok* is planted in front of the *jeu*.

In the following days the *enak tsjim* (bone house) is constructed. In the old days it was made with the bones of beached whales, which even today are found in the area of Flamingo Bay. Today, however, the *enak tsjim* is of wood. Raised on poles, it is about four metres long and two and a half metres high at one end, sloping down to just above standing height at the other. A porch is added to the lower side and over it is hung a carved head of a crocodile, decorated with sago leaves.

The young men enter the *enak tsjim*, their faces blackened with soot. From then on they must not leave the hut except at night. When they

146

do go out a horn is blown so that the women of the village can hide. Up to as many as 25 youths will live in this house for three months or more. Here they are taught the age-old customs, rituals and beliefs of the tribe by their elders. During this time their food is brought to them by family members. No breakfast is allowed; only lunch and dinner.

During the time that the boys live in the initiation house, the men go to the forest to collect sago, catch fish and hunt wild boar. At this time, too, holes are cut in some of the sago palms for the capricorn beetle to lay its eggs.

A *waramon* (spirit ship) is carved in great secrecy in the *jeu*. Each night it is covered with sago fronds, safe from prying eyes both human and spirit. Fashioned out of soft wood, the spirit ship is a bottomless canoe with figures carved on the inside representing the spirits of the rivers and streams. There is also a crab-like creature, which is believed to live at the bottom of the river, and a female figure representing a woman from the village of Jeni. In the centre is carved a large turtle, the symbol of fertility; and at the prow the head of a hornbill, or the figure of a cuscus; also other carvings shaped like ears.

When the carving is completed, lime is sprinkled on the dampened surface and red ochre and black soot applied to the figures. The *waramon* is then decorated with cassowary feathers, wrapped in sago leaves and tied into a bundle with rattan. It is left in the *jeu*, whose entrances are now closed, and all the villagers go into the jungle to collect sago and grubs. On their return, the women remove the rattan securing the entrances of the *jeu* and the men go in and carry the spirit ship to one of the doors, pushing only the prowhead through the opening. After three such thrusts the ship is then carried down from the hut and placed in front of the initiation house.

A young boy, not old enough yet for initiation, enters the *enak tsjim*, followed by an elder who lifts the child up by his ankles and climbs with him through the roof of the hut. Outside the village men pretend to aim at the child with their arrows and spears as he is thrown back to the crowd.

A drummer decorated in paint and feathers leaps up onto the porch of the initiation house and pounds his drum as each youth emerges and steps over the spirit ship, passing semen on it as a sign of manhood. It was explained to me that semen represents the human river of life; for rivers are the semen of the world, the source of all life.

After this the boys lie on the porch slats, facing upwards. Underneath the porch sit elders of the village who, with a razor (in the old days it used to be a clamshell), draw quick cuts across the boys' buttocks, while other men above the boys mark their chests. As he stands up each boy is struck on the back with a banana stalk to make him grow fast like a tree, and again on the genitals to make the male organ grow strong.

With wild whoops and much shouting, leaping and prancing, the boys rejoin their parents who have waited several long months to receive them back into the family.

When I called on the Croiser Fathers and we were discussing the initiation ceremony, they told me that when the young men go to the bone house, they wear a small initiation purse containing ritual objects. The fathers also said that nowadays no initiation is felt by the young men to be complete unless they also receive a blessing and communion from a priest. Truly a meld of *adat* (custom) and Christianity which combined, creates harmony in the Asmat world.

I have described the initiation ceremony from among the many rituals and feasts observed today in Asmat, because it is a prime example of how the old and new can walk hand in hand.

The Phallic Symbol in Asmat Culture

In many cultures the phallic symbol represents male dominance. So it is with the Asmat, who carve the male symbol on their ancestral poles, shields, prows and figures. At one time, to demonstrate dominance over the enemy, the male organ was displayed before an attack.

Perhaps because of a lack of knowledge of its significance in art and local customs, the phallic display is unfortunately regarded by some as pornography.

To quote from Gunter Konrad in the *Asmat Sketch Books* by Bishop A. Sowada, "The unadulterated Asmat culture manifests parental forms of expression observed as phallic impositions and domination."

Today the intricate Asmat laws, beliefs and rituals are poised in delicate balance between the tenth and twentieth centuries. It is to be hoped that they will preserve the best of the old and adopt the best of the new.

After the canoe has been painted in stripes of red and white and sago fronds inserted into holes bored in its upper edges it is then turned upside down. The master bulder of the canoe is given a clamshell with which he traces a long line the length of the charcoaled keel enabling the canoe's spirit to escape.

The 'birth' of a canoe - a canoe-maker hollows out a tree log.

On the Sirets River near the village of Amborep, Asmat men set out in their painted war canoes on a mock headhunting raid. Up until as recently as twenty years ago, the Asmat warriors used to apprehend unwary strangers on the river, truss them up and throw them from canoe to canoe. Women waiting on the river bank would drag the captives to the longhouse where they might be beheaded and eaten, to appease the ancestral spirits of the clan.

Top: Clam shells are burnt and crushed to a powder which is used when making white paint.

Bottom: Inside the traditional *jeu* (men's hut) men converge on the bark floor to carve or chat beside their individual firepits.

Opposite: Outside the hamlet of Agats, Asmat children play in the tidal river near the site of an old *jeu* hut.

On the Casuarina Coast. In the early morning light, Asmat canoers paddle out to sea in search of fish.

In the soft light of the waning day an Asmat woman stands up in her canoe to trace her way through the waterweeds.

As they wait for the fishermen to return, an Asmat mother and child from Bajun or the Casuarina coast build a fire to ward off the morning chill. (refer to page 156)

Fishing nets make good headwear, as this young boy demonstrates. (refer to page 157)

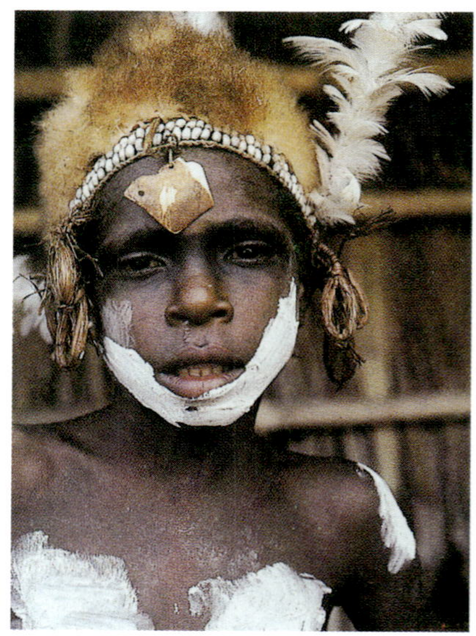

Top left: To enhance his young face a Brazza teenager blackens his eye area with charcoal from the firepit and smears clamshell powder paint, symbolic of strength, on his face and upper body.

Top right: Bedecked with necklaces and wearing a cuscus fur hat trimmed with coix seeds, the son of an Asmat warrior stares haughtily at the intruding visitor.

Bottom: Reminiscent of warring times an Asmat adorned in cuscus fur and cassowary-feathered headwear, with clamshell whitening on his face paddles out in a war canoe on the Sirets River.

Sitting on the verandah of his hut in Per, an Asmat man beats out a tune on his hollowed out tree drum. Pitch is changed by placing lumps of heated beeswax on the lizard skin face of the drum.

Cone-shaped baskets made from sago palm fronds, for holding sago starch.

Returning from the jungle with freshly-gathered sago starch in woven palm carriers, women from the village of Atsj help each other to unload their canoe.

The last catch of the day — sunset finds a sole fisherman on a boat moored outside the town of Agats.→

In the last rays of the sun an Asmat child pulls in his fishing line from the Asewets Ever.

Primitive Art of Irian Jaya

Dani Craft of the Baliem Valley

Unlike the famous Asmat woodcarvers in the swamplands, Dani artisans in the Baliem Valley confine their craftsmanship solely to items of daily use and apparel: adzes, spears, bows and arrows, *noken* (carrying-nets), *thali* (grass skirts), *jogal* (skirts of orchid fibre for married women) and of are, baskets.

Shells, animals' tusks and teeth, birds' feathers and fibres from trees, reeds and other plants are used, mainly as body decorations. The colours — rust, yellow, red, green and white — are obtained from either vegetable or mineral dyes.

Oval stone adzes, once of universal use in the valley for felling trees, trimming wood and butchering pigs, are last being replaced with metal blades. Adzes are made from hard metamorphic rock or chlorite. The stone is sharpened against limestone rocks by the river's edge and bound with fibres to a branch or piece of wood which forms the handle.

Bows are simple and carved from laurel wood. Arrows, tipped with myrtle wood, are more ornate, their barbs elaborately carved. In the days of tribal warfare these barbs were designed to break off on penetration of a body, causing the wound to fester. Flat-tipped arrows are used at pig-killing rituals while pronged arrows are used for hunting birds and occasionally, fish. A geometric design is carved around the base of the arrowhead as well as along the reed shaft. The carver works with little more than a sharpened boar's tusk.

The *suale* is a magic wand of cassowary feathers bound onto a wooden handle with an orchid fibre braid.

Armbands, worn by the men, are of woven bracken fibre. The exterior of the stem is stripped away, leaving two threads of brown and black which are cleaned of the sticky substance which adheres to them and then woven into *teken*.

The *walimo*, a bib of minute *nassa* land snail shells, has a backing of woven bark fibre. The fibre is rolled on the thigh to form a rough thread, more like string, and this is then woven into small strips which are sewn together to form the bib. Lastly, the shells are stitched on.

Most male body ornaments are woven from various plant fibres to which are added feathers, shells and seeds.

The *jarrak* is used as a head decoration on newly born babies and draped over the dead before cremation. Each is a strip of woven fibre three or four metres long, interlaced with cuscus or dog fur as well as tiny colourful feathers and large cowrie shells. Given as gifts and once used to measure the girth of pigs, they have become scarce in the Valley.

Before the great wedding ceremonies held every four or five years, vast quantities of red, yellow and white fibre are collected from wild orchid stems found in moss forests above 1,800 metres. Painstakingly, eight bands of fibre are woven into a braid and huge balls of this braid are set aside for the making of the *jogal* or bridal skirt. The collecting and braiding is strictly men's work, while the actual making up of the *jogal* is women's work.

Je stones are bridal or funeral gifts. They are long, flat, well-polished stones quarried from an ancient site at the Jelime River, and usually a deep green in colour. Small bands of woven orchid fibre, boar's tusks and dried pigs' tails are used to decorate them. On the facing page, ancient *je* stones, traditionally exchanged as gifts at Dani funerals, rest upon the *noken* in which they are wrapped. Bound with the *jarrak*, the stones are kept hidden in the depths of the chief's hut and this was the first time they had been revealed to an outsider. Even the villagers, on this occasion, were forbidden a sight of the *je*.

Noken are made of fibres from one or two trees known locally as *bekelwalet* and *yanoguik*. Colours are derived from vegetable dyes mainly, the most common being yellow (*celleg*), white (*yangguik* in the highlands, or *kuron* in the lowlands), red (*momion*) and blue

Sacred Dani *je* stones (exchange gift totems) lean against the stones' *nokens* (given as an exchange gift at Mulligay's funeral).

(*balakie*). Only the red dye is extracted from a mineral source, a local stone which is powdered and mixed with water.

Noken may be made in the company of others, but the headband (*awirihik*) must be completed in the privacy of the weaver's hut as no man may set eyes on this sacred finish.

Sun-dried fibre is dipped into one of these dyes and then finger-woven across lengths of bamboo strips in a simple design of loops until the *noken* is long enough to hang from head to below the anus in order to prevent marauding ghosts from entering the body.

Today the women follow the same rituals, but the fibre used is more often brightly coloured imported thread purchased at Wamena market.

Nowadays well designed baskets woven out of plant fibre and decorated with coloured strips of red, yellow and brown orchid fibre are seen in the market place. Made in the southwestern valley of Tiom as well as at Anggruk in the highlands above Wamena, they are a fine example of how quickly the Dani tribesmen have adapted to new ideas and the opportunities brought by tourism.

Asmat Woodcarving

In these pages I merely brush the surface of the intricate art and hidden symbolism of Asmat woodcarvings. Many hooks have been written on Asmat art forms and the religious rituals pertaining to the carvings. The great collections housed in museums in New York and Europe preserve forever artefacts carved by hands long since united with those of their ancestors.

Today even the carver himself does not know or understand all the hidden meanings in his work. His skill is directed as much by tradition as by a quest for beauty; for Asmat art is the outward expression of ancient beliefs and customs. In the past, carved figures were named after slain ancestors inhabiting the spirit world of Safan, and were a reminder to the living that the dead required revenge. The ever-present phallic symbol was to intimidate and express dominance over the enemy.

Every Asmat male is a carver in his own right, but the prestigious title of *wow ipit* is bestowed on only the greatest among them. These master carvers are commissioned by the villagers to undertake special carvings such as ancestral poles. The *wow ipit* is rewarded with food for himself and his family, including such delicacies as sago grubs.

Measuring up to ten metres high by 50 centimetres in diameter, the *bisj* or ancestral pole can take months to complete. Ancient rituals are closely observed from the time of selecting a suitable mangrove tree to the last finishing touches.

The *bisj* is in three parts: a tapering foot which enables the pole to be inserted easily into the ground, a tall trunk on which ancestral figures are carved standing one above the other, and a flying buttress stylized with headhunting symbols.

On completion of the carving, a ceremony is held when the orifices are added to each figure, which is named after an ancestor. The pole is then whitewashed and the figures etched in red and black paint.

The final ceremony takes place when the *bisj* is erected in the ground outside the men's hut. A mock war is enacted by the men, followed by the women dancing and shouting to frighten away the souls of the ancestors represented in the pole.

Carvers on the coast, including the Casuarina Coast, and inland specialise in *bisj* and sago grub trays; northwest carvers specialise in spirit ships and woven masks. However, all the Asmat tribes carve figures, food and paint bowls, drums, paddles, prowheads, shields and horns. Shield designs vary from region to region.

Softwoods are used for ancestral poles and spirit ships which after use, are returned to the forest and left to decompose. It is believed that the spirits within the carvings will protect the sago grounds.

Ironwood, which is extremely hard, is used mainly for figures. The wood has a pinkish hue which later turns dark brown.

Shields carved in one piece, including the handles, are made from the upper roots of forest mangrove trees. (Interestingly, the same mangrove roots are used as a forest telegraph by the nomads of the Upper Brazza region.) These same trees are likened as human beings since when they are felled their sap runs blood red.

The nibung palm is used for making paddles, spears and drums, and a variety of woods for bowls and other artefacts.

Before the advent of metal in the 1950s, cassowary bones and claws, along with clam shells and boars' tusks were the carver's tools. Today flattened nails and modern metal chisels have replaced the cassowary bone. However, clamshells are still used for scraping and polishing, and in some areas the cassowary claw remains the preferred tool for etching.

Throughout Irian Jaya, three basic colours are used: white, red and black. For the Asmat these hold magical qualities. White (*mihi*) comes from baking clam shells in between layers of firewood. The ash, crushed and often mixed with cockatoo feathers and crab shells, produces a pure white powder. In regions such as Brazza, where there are no shells, clay is used.

Red (*wasah*) comes from mud found upstream and often traded downstream for sago and fish. The clay is baked slowly over a fire until the desired shade of red is reached. It is then stored in bamboo containers until needed. Red is a favourite colour, often worn around the eyes to give the wearer a fierce appearance like that of the black king cockatoo. Red stripes painted on a canoe are said to make it go faster.

Black (*sosok*) is made from crushed charcoal. Among other things, it is used for the hair and groin of carved figures. The men mix it with smoked lizard fat and paint it around the eyes to make themselves more desirable to the women.

When working on the wood, the carvers take great pains to protect it from the sun. Carving is done either on the verandahs of their huts or inside the *jeu*. If it has to be done outdoors, as in the building of a canoe, a canopy of nipah palm is constructed to protect the wood.

In recent years Asmat art has taken on many changes. Old carvers still work with ancient symbols known only to themselves, but, less restrained by tradition, the young have introduced extensions to the old designs and have gradually evolved their own styles and forms. The Asmat remain the most skilled carvers of all primitive art and their work is prized by collectors.

To stop headhunting and warfare, the carving of ritual objects was forbidden in the early sixties, and many precious artefacts would have been lost to the world had it not been for the foresight of the Croiser Fathers who collected the artefacts and encouraged the Asmat to preserve their artistic skills.

The late Michael C. Rockefeller visited Asmat in 1961 to record and preserve the art of what then appeared to be a disappearing culture. His fantastic collection is currently housed at the Museum of Art in New York, silent figures from a lost domain.

On 17th August (Indonesia's Independence Day) 1973 the Asmat Museum of Culture and Progress was opened It houses over one thousand artefacts, some extremely rare, all beautifully crafted. While built primarily for the preservation and display of Asmat art, the Museum also has a collection of artefacts from the Brazza River region, donated by Dr Gunther and Ursula Konrad. There is an auditorium for lectures, films and meetings. Once a year Bishop Alphonse Sowada of Agats holds a competition for Asmat carvers and the best pieces are displayed in the auditorium.

At Sawa-Erma, a village on the banks of the Pomats and Wassar Rivers in the northwest, stands a unique church constructed by the villagers. I was invited to view it by Father Vincent, the resident missionary. It resembles a traditional L-shaped long house. Each family in the village has its own fireplace, decorated with fibres from the paper mulberry tree, and on Sundays the family gathers here to pray. Throughout the church are miniature ancestor poles carved with detailed ancestral figures painted in traditional colours.

On a raised platform is the main altar, also with its own fireplace and surrounded by beautiful carvings. The walls of the church are decorated with large murals depicting the new Asmat way of life; yet included in these paintings are the figures of the spirits of trees, birds, fish and animals. Again, here is harmonious transition—a blending of two worlds two beliefs in this fine example of Asmat art.

1 *Je* stones traditionally used as exchange gifts.

2 *Walimo* — man's collar of snail shells (*nassa*) sewn onto a bark backing.

3 Stone adze of metamorphic epidote and chlorite rock alongside a decorated finger-joint cutter.

4 *Mikak* — bailer shell necklace.

5 A rare amulet of caterpillar cocoons fastened to a finger-woven fibre cage. The cage holds a nut decorated with tiny bird feathers.

6 Arrows decorated with orchid fibre, used for killing pigs and small birds.

7 This *suale* — a magic wand of cassowary feathers bound to a woven orchid fibre handle — is used at initiation ceremonies.

8 *Jarrak* (cowrie shell bands), now given at exchange ceremonies, were once used as pig-counters or 'tape measures'.

Shoulder bag, decorated with feathers from tiny red and green parrots and bobbles of dog fur.

Dani man's amulet bag decorated with pigs' tails and tusks.

An assortment of woven baskets and armbands beside a *noken.* →

Symbolic of the Asmat warrior's strength and power, shields once used in tribal warfare were believed to have magical powers which protected their owners from the enemy. Carved into the shape of a phallus, a shield would be decorated with designs of flying foxes, cassowaries, praying mantises or cuscus tails. Figures of the cockatoo and hornbill, symbolic of headhunting, were also used.

This miniature shield with its intricate design was made in the Sawa Erma area as a collector's item. →

174

Drums are carved from a single log into hourglass shapes which are then hollowed out with the use of hot embers and a digging stick. Designs are etched into the body, with the handle usually shaped in the figure of a man or animal.

The lizard skin covering the top of the drum is sealed with lime and human blood scraped from the carver's leg with a clamshell. The skin is held in place with a band of rattan which is tightened over a fire; tuning is done by adding three drops of beeswax to the skin's surface and warming it in the fire.

Drums, named after dead people, are played at ceremonial feasts and when not in use are stored in the rack above the fireplace in the longhouse.

Bowls used for holding sago grubs or for mixing paint were crafted with figures facing uppermost, to enable the user to look at the ancestral face represented in the carving. Later, when the bowls were sought after by collectors, the figures were carved facing outward. →

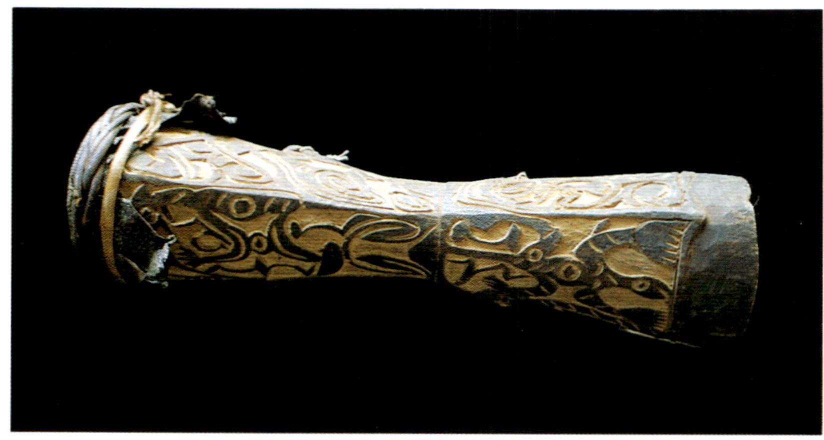

Carved figures represent tangible portraits of dead ancestors. The *wowipit* (master carvers) were often commissioned to carve the figures before great feasts, after which many of these figures were discarded.

Today's *wowipit* continue to carve the traditional entwined figures, but have also added a new dimension to their art — single figures depicting various aspects of Asmat life. →

178

Partially hidden by grass, old weathered figures carved from a single tree stump stand guard outside the *je* (men's longhouse) in the village of Jaosakor. These symbolic figures were often displayed at the entrance to the *je* hut in order to protect the inhabitants.

Smaller poles were often placed inside *je* huts. Here in the hamlet of Sawa Erma the villagers carved traditional poles and erected them in their church which they fashioned after the style of the men's hut with firepits for individual families, the main firepit representing the altar.

Ancestor and trophy skulls are things of the past. Trophy skulls were taken from enemies and the brains removed through a small hole and placed in a ceremonial bowl. This was then presented to elderly men and women to give them strength. The jawbones were tossed to the women who used them as centrepieces for their necklaces, while the skulls were hung in the doorways of huts, or in clusters inside the *jeu*.

Ancestor skulls became a part of family life. Worn on the backs or chests of men, or used as pillows, they were a constant reminder of the need to keep in contact with the ancestors. The skulls were decorated with grey coix seeds, red abrus seeds and cassowary feathers.

Bone daggers used in battle were made from the jawbone of a crocodile or the shin bone of a cassowary bird. The hilts were trimmed with woven bark string, coix seeds and cassowary feathers.

Masks are made in great
secrecy in the *jeu* by the
men. Fibre from the dried
bark of the paper mulberry
tree is rolled into string on
the ball of the weaver's foot,
then woven into the desired
shape. The mask is painted
white and decorated with
cockatoo feathers; ears and
eye-pieces made from shells
give a frightening appearance
to the mask. Sleeves and
skirts of dried sago fronds are
then stitched on to the mask.

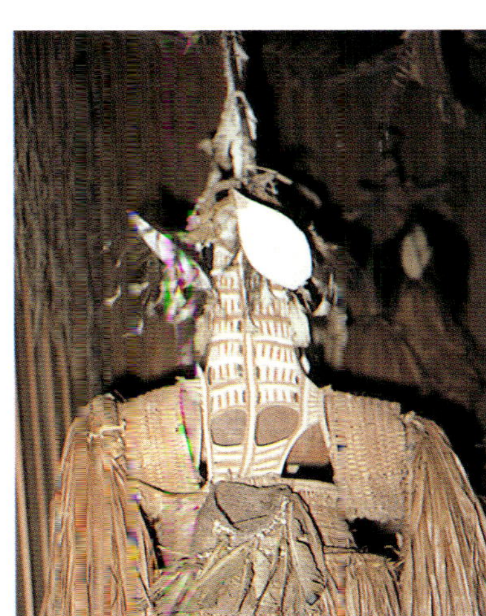

Bags used by men and women to carry sago starch, firewood and
smaller objects are made from dried sago fronds and decorated
with feathers, seeds and paint.

Smoking pipes of etched bamboo used by the men.

Initiation bags of mulberry tree bark dyed red and white, in
which are kept sacred objects, are worn around the necks of
young boys whilst they live in the *emak tsjm* (bone house).

Nosepiece carved from pig bone.

Headbands, worn only by the men, are made from cuscus fur and woven sago fibres.

Women wear skirts of sago fibre, with the long strands drawn up between the legs and knotted at the back.

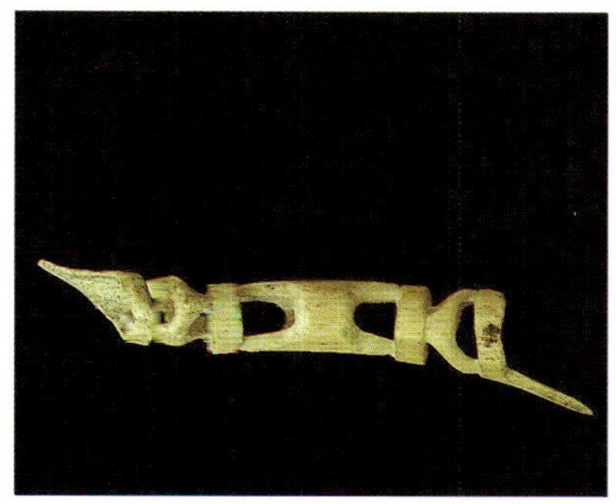

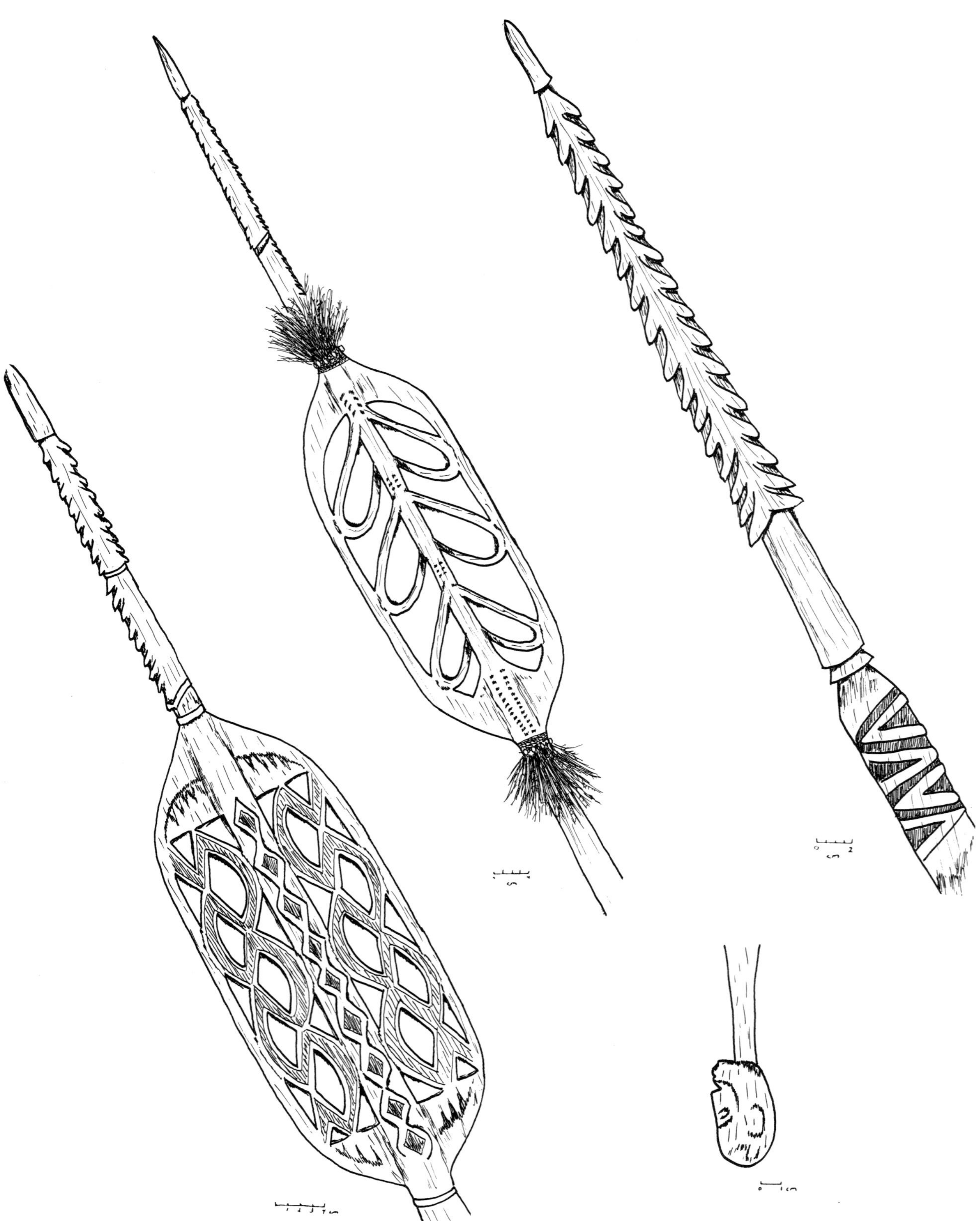

←Ornately carved, spears are often tipped
with a cassowary claw and trimmed with
cassowary feathers. Designs most commonly
found on them are those of the praying mantis
and cuscus tails.

Arrows are generally less ornate and are used to hunt
wild pig, marsupials, birds and fish.

1 This praying mantis design, symbolic of warriors and headhunters, is usually carved on spears and shields.

2 Hand design engraved on shields and *bisj* poles is a reminder of the presence of ancestors throughout Asmat life.

3 This animal spine design representing a warrior's strength and power is found mostly on shields.

4 Carvings of turtles on shields and prowheads of canoes are symbolic of fertility since turtles are known for laying many eggs.

5 The water snake design indicates a commander's strength, cunning and agility, and is usually engraved on spears and oars.

6 The hornbill beak design is most commonly found on shields, drums and prowheads. A hornbill is an emblem of a head warrior.

7 The turtle design is often found carved on the tops of shields that come from the western Asmat region

8 Pari fish design, symbolising protection

9 Ancestor figures are often carved on the tops of shields that come from the Pirien and Omandesep areas. The shields are named for the ancestor represented by the figure.

Paddles of great length, with carved blades and feathered poles, are used by the men to manoeuvre their canoes. Standing in an upright position, the canoeist bends on one knee as he digs the blade into the river.

189

Land of Many Facets

Flora

It is impossible to express through words or pictures the true magnificence of the many facets of Irian Jaya's landscape — her lofty mountain peaks, desolate alpine meadows, gentle upland valleys, azure lakes, rivers of rushing white water as well as sluggish silt, dense rainforests, swamplands and crystal white beaches.

The jewels of the land are the flowers gracing the river banks, colouring the savannah, crowning the forest, and forming miniature floral bouquets in an otherwise barren alpine landscape.

The land is both placid and hostile. It sleeps for decades, then without warning an earthquake rends the earth asunder and sends mountain slopes crashing down.

There are four distinct zones: the lowland swamps and rainforest, the savannah, the moss forests and the alpine regions, each with its own unique appeal. From each region the natives cull plants, shrubs and trees for medicine, clothing, building materials, utensils and weapons. However, they take from the land only what they need to survive.

The Rainforest and Lowland Swamps

As I walk in the heart of the rainforest the world around me is in its primaeval state. Its ancient innocence clears the mind and sharpens the senses. The plants around me are so swollen with life that I seem to hear them sipping the moisture laden air and see them growing.

The early morning orchestra of birds and insects is replaced by the uncanny silence of the noonday heat. A tree crashes to the ground, no vines to cradle its fall, and the whole forest gasps at the intrusion of noise.

Jewelled butterflies dart silently from plant to plant. Eve's reddest apple would pall beside the vivid hues of the strange forest fruit clinging to the tree trunks. Blossoms of red, orange and yellow - tangible fragments of yesterday's sunset - lie upon the oozing black humus floor of the forest. Ferns, shoulder high, herald the proximity of water, one of the myriad streams that, seen from the air, carve silver patterns below the green canopy.

Our progress is slowed by the rattan creeper, deceptively innocent in appearance until it tears at flesh, hair and clothes, or whiplashes the unwary to a standstill. Tightly entwined creepers form a natural curtain, obscuring from view the barely discernible trail. They grapple their way to the crowns of the giant trees in search of sunlight, while huge screw-pine epiphytes, tree orchids and birds' nest ferns take up squatters' rights in the forked branches. A strangling fig triumphantly ensnares its host in a death hold.

Nearby the enormous buttresses of the mangrove tree dominate all lesser trees — the ironwood and softwoods which the natives use to make houses, canoes, spears and shields. Beneath my mud-clad boots silt and roots vie for space in the peat swamp, combining with thickets of spiny palms to make passage through the forest barely possible.

The unwary insect is food for the *Nepenthes ampullaria*, commonly known as pitcher plants, whose cups hold sweet water for the thirsty and delight the eye with their delicate designs, hair-like fibres and soft green colours. Beyond, red and silver enhance the dark green feathery leaves of ground herbs growing at the base of ginger. Stalks of wild banana stand sandwiched amongst the lesser-known vegetation.

Thriving in the dark humus, fungi of woodland colours and every shape and size make their homes on the rotting branches and tree trunks littering the forest floor. They create the perfect environment for thousands of insects as well as slinky, quick-moving leeches.

Standing in the forest, I am actually aware of the abundance of natural resources yet untapped, and overwhelmed by the strength of life surrounding me. This is indeed a life-sustaining land, not only to flora

and fauna but also to the nomadic tribes which make this region their home.

The Savannahs

The valley of the Baliem River is a wide swath of grassland, while the slopes of the surrounding mountains are sporadically covered with flowering chestnut trees, myrtle, laurel and *araucaria*, an ancient tree whose upward-curving branches hold clusters of needles.

Among the trees are giant rhododendrons, their large white and orange blossoms startling against the dark green leaves. This region is home to ground orchids, some with purple flowers and yellow stalks, others with white blossoms. The fibrous stems are woven into braid for the Dani *jogal* or bridal skirt.

Below the mountain slopes feathery-needled *Casuarene subelatem* border the rivers and streams of the highland valley. There are small copses of beech, their trunks and branches adorned with liverwort and garlands of silver fungi.

Early one morning Wynoco and Siba asked me to join them for a walk along the Akei River and out into the surrounding countryside. As we walked they pointed out many shrubs and plants, explaining the use of each. Since most of the names given to the various plants and trees were in Dani language I had to rely upon memory and sketches for identification.

On the sides of a limestone outcrop considered sacred by the local people, were miniature white and brown orchids, sprays of orange ginger blossoms, vetch and lupins. Yellow jonquil-like flowers hid amongst the profusion of ferns in the rock crevices and lining the river's edge.

Where the tall sedge grass dressed the land, clumps of rhododendrons stood in isolation, covered with trumpet-like flowers of white and yellow. There were also thick clusters of *aquilaria* bush, whose fibres are used to make the Dani *nokens* or net bags, and hidden in these bushes were the nests of tiny birds, like wrens but coloured white, yellow and brown.

Bracken dominated the pathway alongside the river. These the men strip, using the inner fibre for armbands. A small quail nearly flew into our faces as we passed, fleeing its ground cover of cane reeds.

Crossing over a stream we came to a brush bog. The tannin pools held waterweeds collected for the children's skirts. Dragonflies and waterbugs skimmed the surface, watched with interest from a rock by a frilled puna lizard with hooded eye and darting tongue.

As we progressed my head began to spin with the names my companions reeled off to me, and the uses to which they put each plant, bush and tree. The Dani have over 300 names for plants and by the age of ten a young child has learned most of them by accompanying his elders through the countryside on their way to hunt, fell trees, or on visits to other villages.

We decided to turn back as the sun had crept high in the sky. Shielding our eyes against its brilliance, we saw hawks circling in search of small rodents. Before we entered the woods on our way back, we quenched our thirst on red, sweet-sour berries from wild strawberry vines which bore fruit and flower simultaneously.

In the damper areas of the wood we found mushrooms and fungi of many varieties, three of which, I was told, are taboo to the Dani. More than likely they were poisonous, since the Dani eat most of the other varieties.

Close to the village we stopped to watch an enormous spider spinning its web between two branches. Its cobwebs are often collected, greased with pig's fat and worn around the neck as an amulet. Sometimes the giant spiders are caught and placed in the pig's hut where they continue to spin their webs conveniently for the Dani to weave into amulets.

I learnt many names on that morning walk, as well as an appreciation for the Dani's amazing knowledge of medicinal plants and of their genuine concerns for the land which for centuries has been their domain.

Moss Forest and Alpine Valley

High above the Baliem Valley in the Southwestern region lie the forests of gnarled oak, pandanus, conifers and deciduous *Diorycarpus*, their feet smothered in filmy ferns and their heads in the low clouds. Also to be found here are the Rata of New Zealand fame, with spiked red blossoms on spindly branches. On the slopes of Mount Hundepma the rhododendrons bear tubular-shaped pink and red blossoms which are pollinated by honeyeaters, unlike their sister species in the Valley whose yellow, orange and white blossoms are pollinated by moths and butterflies.

Spagnum mosses hang in curtains from twisted branches. Other branches are hosts to woody epiphytes, eerie in appearance. Underfoot, bog mosses silence our tread. Among sprawling tree roots, orchids of fingernail size and delicate colouring hide their miniature profiles. Here I discover a tiny black orchid - a very rare find apparently, judging by the exclamations of the botanists accompanying me.

I am with a group from the National Council for the Preservation and Conservation of Plants, all botanists, all peering excitedly into, under and above everything, rattling off Latin names as foreign to my ears as any native names, collecting leaf samples and stooping to take photographs. They tell me that they think many of the plants have yet to be catalogued, and we envision with delight the surprise on the faces of those yet to hear of our finds.

Beyond Andelikmo, by the side of the waterfalls we find *ostilbe* with branches of pink flowers and the red-barked *leptospermum* tree, usually associated with Chile and New Zealand. As we continue our journey, the moss forests eventually give way at Yaboryema Pass to a black, boggy terrain. Now we are in a miniature world of heath and alpine plants. I am able to recognise some familiar flowers among the myriad that are new.

A blue gentian surrounded by heath grass is mirrored in a dark brown pool. Dwarf yellow jonquils, cotoneaster and a yellow-berried alpine plant catch my camera's eye. *Sisyrinchium*, an ankle-high plant with bright red trumpet flowers, tiny rhododendrons and potentilla, bright yellow celandine, carpet the ground around me, and I am reminded of springtime in the Scottish highlands.

Everyone is shouting out names of plants amid "oohs" and "ahs". Even our porters are infected with excitement and run around hunting for new specimens to show us. Mount Hiddipma is the physical and botanical climax of our expedition. It is time for us to descend this peak and move on to the lake which, we understand, is yet another three hours away. We have been walking now for five days and camping in the forests in tents, stopping each evening and moving on at daybreak.

Descending the slope we find ourselves at the head of a long glacial valley. But our excitement is not for the strange *dickonia* trees dotting the landscape, but for the view ahead of us. In the distance, Peak Trikora (4,500 metres approximately and one of the highest peaks in Irian Jaya) greets us and I for one am speechless. The face of the mountain dips and curves with geological formations shaped thousands of years ago, while below the mountain, at the far end of the valley, Lake Habbema glistens in the misty early morning sunshine. We stop in awe and even our porters are quiet. Here nothing moves; silence reigns. It is as if we have stumbled onto an ancient world never visited by human beings.

Ever since I had seen a photograph of this lake in Robert Miton's *The Lost World of Irian Jaya*, I had dreamt of seeing it in real life. At last, thanks to the botanical group, dream and reality are fused.

Spell broken, we click our cameras and begin our journey through the valley. The wet ground gurgles underfoot and we sink into bog pools. The landscape is strange - reddish, spiked bog grasses, stunted trees and unusual plants. *Dickonia* trees huddle at the head of underground rivers which surface a little way, only to burrow underground again, leaving behind an echo of their subterranean streams.

Tiny plants and ferns find refuge among the round glacier-pitted rocks; gnarled and stunted bushes bow their heads against the onslaught of alpine winds. Gone are the colourful birds and forest. As we walk we disturb a few brush birds of subtle colours and a small quail darts across our path.

The bogs dry out as we approach the lake, giving way to a surface of peat and entwined roots. Our porters have gone on ahead, leaving us with only a few Dani who came up with us from the Baliem Valley.

Wynoco, like the others, is cold and as we near the lake we find that the porters have already cut some low conifers and turned them into crude huts which they are covering with ferns and roofing with black peat sod.

The lake lies at an altitude of 3,700 metres and although the sun is shining, it is windy and cold. Over all there is a feeling of eeriness, as though there are others watching us.

Leaving the porters to their work, I walk down to the lakeside just beyond our campsite. The edge of the lake is full of weeds and I am surprised to see coots with black bodies and white-crowned beaks swimming and diving in the water. There appear to be no fish and this was confirmed later when the men went fishing and did not get a single bite.

On an exploratory walk just above the lake, away from the campsite, we find many stunted conifers; attached are enormous woody epiphytes with small green leaves growing at the top so that they look like quaint pineapples. I find one on the ground, broken open. Inside, it is like a yellow honeycomb covered with thousands of black ants. The honeycomb does not taste sweet, so I leave it for the ants.

The ground is quilted in places with thick star-shaped yellow and red moss and clumps of spiky grass. Even the tree trunks have moss on them, six or seven centimetres in length and mostly a deep red.

Struggling up and down the difficult terrain, we find many more unnamed plants and grasses. Around a bend we see that there is still much more of the lake yet to be explored, and wish we had a boat. But we are hungry and tired, so we return to the camp.

For the last two days there have been no villages; the only people in this part of the world are our group of eight, Rudi, Wynoco and twenty porters. The porters have run out of food and are dependent on what they can get from these villages. They are eagerly awaiting the return of a runner with sweet potatoes.

Returning to camp we see a huddle of men and find that someone from far away has heard of us. He has come with a bundle of freshwater crayfish, very much alive, which he wants to sell. Elly, our cook, counts them out and ends up buying them all for what appears to be a song — fifty for five thousand rupiah (about $3). This will be dinner for us tonight, but now for lunch we have just hot soup and a drink. The sky is getting dark and a wind is blowing up, soon to bring rain.

Our time at Lake Habbema was an experience all of us will never forget — the stillness and solitude, and yet the strange feeling of being watched. The nearest I know to this is being on the wild Scottish moors in the Hebrides.

Working, writing, eating, awakening, we faced the dominant mountain walls. Morning mists gave way to sunshine and brilliant blue skies reflected on the surface of the lake. In the afternoons rain clouds drifted up from the forests below, sending us scurrying for shelter, photography and plant collecting stopped for the day. From our dripping tent flaps we could see the porters' huts steaming from the campfires within, giving the campsite a hamlet-like quality.

When the rains dispersed and the moon came out, heaven seemed just an arm's length away. We felt that we had only to reach up to touch the stars. The lake shivered in a nightgown of silver gossamer and the mountains, bathed in moonlight, seem to cut us off from the rest of the world.

On the return journey we took a different route to avoid the dangerous waterfalls we had ascended on our way to the lake. The rocks had been extremely slippery and the way had been fraught with difficulties. Part of the way, the porters had to cut steps in the mountain slope and in doing this they uncovered a path of wooden laths. I wondered, as we slithered and slid and picked ourselves up, who might have journeyed there before us.

Trekking through the forest again, we spent many hours discovering and poring over more new plants, mosses, ferns and flowers, each of us dragging out the time, loath to leave.

Finally, to the accompaniment of birdsong, our porters' tribal chants and pouring rain, we emerged from the forest. Another four days or so saw us back in the familiar Baliem Valley after one of the most exciting trips I have ever made in Irian Jaya.

Fauna

Unlike Kalimantan and other islands of the Indonesian Archipelago, Irian Jaya has few large animals. But what it lacks in these, it makes up for with its hundreds of colourful birds and myriads of butterflies, moths and other insects.

The regions covered in this book are home to the wild pig, which is hunted in the forests for meat. Tree kangaroos, cuscus and other members of the opossum family are also hunted, usually for their fur. Their golden and brown coats are turned into headdresses by both mountain and lowland tribes.

Rats and many other small marsupials are eagerly shot with arrows and then roasted over open fires or used for magical purposes. Flying foxes and bats abound throughout the high forests surrounding the Baliem Valley, but it is the nomadic tribesmen, not the Dani, who use the bat's wing-tip bone as a nasal ornament.

Cassowaries, large flightless birds, scuttle through the forest building enormous nests of forest litter and leaves in which to lay their blue-green eggs. When found, these are eaten and the shells decorated with Asmat designs. Hunted with bow and arrow, the cassowary's flesh and feathers provide food and decoration. Dani men turn its long silken feathers into sacred wands, while in the past the Asmat people made cassowary thigh bones into daggers, and claws into arrow tips as well as etching tools.

Exotic birds such as the bird of paradise are to be found in remote highland areas. They can be observed performing their wonderful acrobatic tricks at daybreak and in the evenings. Dwellers in deep foliage, they play and sing elusively in the high trees. The male, with impressive silken flowing tail and resplendent coat of emerald green and orange, tinged with golden feathers, once graced many of the forests throughout Irian Jaya, but has now been hunted almost to extinction. The female, with her drab brown colouring, is left to fend alone and now inhabits less frequented areas with another mate.

Parrots, lories and cockatoos are found in the trees of the lower secondary forests, along with honeyeaters, kingfishers anti tiny finches.

Primary forests echo to the screech of black palm and sulphur-crested cockatoos. Drumming hornbills disturb the treetops while dusty blue Victoria pigeons brighten the jungle floor or perch, camouflaged against the sky in the upper branches.

In Asmat art black-faced fruit-eating birds such as the hornbill and the palm cockatoo have headhunting connotations. The plumage of many birds long survive their fragile bodies, living on as elegant headdresses for the tribal people.

Birdsong in the mornings is replaced by the whine of cicadas as the day grows hotter. Stink bugs, as the native children call them, are tasty morsels caught by quick little fingers. Plucked buzzing from a tree trunk, they are nimbly rolled in a leaf and popped into the mouth, their crunchy texture and sour tang definitely an acquired taste.

Dancing through the grass alongside a burbling river or a quiet, pensive stream are butterflies of delicate and vibrant hue. Pollinating the valley flowers, they colour the land when the birds take shelter from the sun's rays. Darkness brings out moths and other insects, their delicate wings shimmering like strands of silk in the night, while riverside bushes turn into a magical fairyland, glowing with pinpoints of light that twinkle on and off from the thousands of lightning bugs. Raucous frogs strike up an ear-splitting chorus.

Along the coast at low tide skipjacks play tag in the oozing mud while small crabs with blue-green shells rush in and out of their hide-outs. A solitary kingfisher and a wading heron watch for the opportune moment to swoop, leaving an empty space where an unwary crustacean once raised its head.

The tide rolls back in, bringing food for man in the form of fish both large and small, but sharks and whales stay well beyond the prowling canoes. Occasionally the primaeval crocodile basks on a log, or swims into a side river in search of prey.

On the shore of the Arafura Sea flocks of pelicans rest like miniature beached ships. Long ago, taking refuge from an Australian drought, the pelican took up a residence on Asmat shores. Since then the people of the swamplands have immortalised the bird in their wood carvings.

A precarious balance is maintained between primitive man and primitive fauna. As one changes, so will the other and I wonder if the grandchildren of the Dani, the nomads and the Asmat will thrill to the brilliance of the bird of paradise or the drumming of the hornbill.

Birds to Paradise (*Parasidia apoda*), sometimes referred to as the 'birds of the gods', are known for their long hooped and colorful tail feathers which are in full plumage in late August. Females of the species are less colorful and have regular tail feathers.

The Victoria crowned pigeon (*Goura victoria*) roosts in trees, but forages for food on the ground. The male's deep booming call alerting its mate echoes through the rainforest. The male pigeons are distinguishable by the white and dark blue plumes of its crown.

Top: The black-capped lori (*Lorius lory*) mostly found in mountainous regions, is often hunted for its feathers which the Dani tribes use in their decorative headdresses.

Bottom: The Papua talyathyina (local name) is found in cloud forests as well as the lower forest regions.

Opposite: Palm cockatoo (*Probosciger atterimus*). Asmat carving of palm cockatoos and their black-headed fruit-eating bird were symbolic in their headhunting raids.

Top: Sulphur crested cockatoo (*Cacatua galerito*) whose feathers are used as body ornaments and for decorating woven sago-fibre bags, Asmat paddles and various carvings, travels in flocks.

Bottom: This hornbill (*Bucerotidae*) of the lowland rainforests feeds on fruits, mice, lizards, and other small rodents. Hornbill motifs engraved on Asmat shields, carving and the prows of canoes are symbolic of headhunting. Young hornbills' beaks are often used as penis shields by nomadic tribesmen.

The jungle chicken's eggs dwaft an average-sized chicken egg. One large egg made a delicious omelette for three hungry travellers trekking through the rainforest.

Bottom: The cassowary (*Casuarius casuarius*), a large, ground-dwelling bird, lives in the forests of Irian Jaya. Its powerful claws are used as a defense weapon.

Opposite: Cuscus (*Phalanger maculatus*), a small furry nocturnal marsupial, which lives in trees in the rainforest is often hunted for meat. their golden or black spotted pelts are worn as headwear by both mountain and swamp tribesmen.

← Tree kangaroo, barely
discernible in bamboo thicket.

Wynoto slashes a trail for us
through the dense undergrowth
of the Upper Brazza rainforest.

The New Guinea creeper,
Muerna bennettii, overhangs a
river. Locally known as Flame of
the Jungle, it bursts into flower
in late July, the brilliant blooms
festooning the riverbanks with
scarlet streamers.

A rare ray of sunlight
penetrating the dense vegetation
highlights a silver-streaked
forest herb.

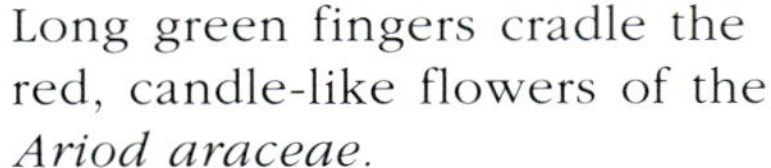

Long green fingers cradle the red, candle-like flowers of the *Ariod araceae*.

Spiny rattan as well as the dreaded sago palm thorns deter men and animals from travelling in the jungle.

Forest lianas span the few open spaces of the jungle, seeking the sunlight of the canopy.

Irian Jaya's rainforests hold a wealth of orchids.

212

Wild flowers of the Baliem
Valley dazzle the onlooker
with their bright blooms.

1 High in the cloud forest, woody epiphytes clasp their roots round the tree branches. Unlike the parasites and creepers, the epiphyte takes no nourishment from its host.

2 Spiky red flowers of the rata (*Metrosideros tremuloides*) bloom in sun-lit clearings in the cloud forest, 2,100 metres above sea level.

3 Early morning light gives the cloud forest an ethereal quality.

214

4 - 7 Miniature orchids found in the cloud forest above 2,700 metres.

8 Reminiscent of candy canes, this ginger is found only in the depths of the cloud forest 2,700 metres above sea level.

9 Brilliant red flowers of the mountain rhododendron.

A double myrmecophyte (plant associated with ants)
turns an otherwise normal-looking tree into an object
of fantasy.

The honeycombed interior of a fallen myrmecophyte
becomes a feast for thousands of ants, who make it
their home as well.

Delicate blue gentians growing on a tuft of bog grass, in a mountain-reflecting pool.

A brilliant carpet of mosses at the foot of the coniferous trees, near the edge of Lake Habbema.

Dicksona trees grow at the head of the glacial valley where Lake Habbema lies. Often mistaken for palms, these trees bear rough-textured, fernlike branches with white undersides.

Bright, bell-like flowers of the *Sisyrinchium* dot the alpine landscape.

217

Change

Change is inevitable, but it is obvious the primitive tribesmen living in the remote areas of the rainforests and swamps of Irian Jaya could not have envisioned the drama that would encompass their daily lives as the industrial world swept into their tribal lands.

The Dani people of the Baliem Valley seemed to have adjusted more easily to their new environment compared to the more remote tribes. They have seen the significant transformation of their world with the implementation of new roads, animal husbandry and new crop cultivation replacing their traditional farming techniques.

This new world encroaching upon Irian Jaya and its people, at a speed unmatched in their previous timeless world, is forcing them to adapt to the techniques of industrialisation, at times without the understanding of the impact it will have on their lifestyles.

New material wonders, tempt both old and young and, with the zeal to acquire these goods, tribal ancient laws and beliefs are undergoing a re-evaluation.

Village chiefs share their responsibilities with deputy chiefs who, chosen by the local Indonesian government, act as the tribe's delegates.

Tribal chiefs, chosen for bravery, kinship or wealth, are still elected by the people to preside over traditional laws, rituals and taboos. They oversee the welfare of the immediate clan. Whereas government-appointed chiefs are responsible for reporting and recording the clan's daily activities as well as acting as a liaison between the tribe and governing officials.

Many of the Dani people are experimenting with rice growing, and the raising of cows, sheep and goats. However these crops and animals are not as yet included in the lifestyle of the average Dani village, but are raised for sale to the newcomers in the area.

Since most outsiders moving into the Baliem Valley are rice consumers there is a need to transform ancient sweet potato gardens into rice paddies. The valley floor is extremely fertile with a high water table, an ideal climate for rice production, but this new crop, which provides for those of a rice culture, depletes the cultivation of sweet potato gardens.

Many Dani villages are moving to higher altitudes where traditional farming using irrigational canals cannot flourish. High on the mountain slopes, the Dani is dependent on the rainfall for the development of their sweet potato gardens.

One wonders how three such diversified cultures as Islam, Christianity and Animism can live together. At present it appears to be working, but with the influx of more outsiders into Irian Jaya one might question whether there will be sufficient leeway for a culture based on pig wealth, ancient beliefs and rituals to exist harmoniously alongside a culture that censures the raising and eating of pork.

Ancient cultures the world over have succumbed to the pressures of unrelated diseases brought into primitive areas. Nowhere is this more apparent than among the swamp and nomadic tribesmen. Though there are modern medicines to help combat the new infections, many tribesmen have little access to them nor are they able to afford them. Most are reluctant to exchange the herbal remedies which they have used for centuries for the unknown qualities of modern medicines.

In many areas throughout Irian Jaya, more tribesmen are attending schools and learning new techniques for everyday living. They are being coaxed into trying new crafts and their youngsters are seeking new friends amongst the newcomers.

Wamena market in the Baliem Valley is full of new vegetables, grown by the Dani mainly for the new towns developing in the Puncak Jaya area of the great copper mine as well as for the hundreds of newcomers to Wamena township. The sale of this produce provides income for the local people who spend it on clothing, seeds, educational materials as well as household goods.

The face of the great valley is slowly but surely developing into an affluent frontier region producing an abundance of varied crops and livestock for a multi-culture population.

One must travel far to find solitude, tranquility, adventure, and villages still maintaining their ancient lifestyles in the small valleys of the mountain regions or deep in the slowly diminishing rainforests of Irian Jaya. Nonetheless they do survive, balancing their frail survival on the edge of the encroaching tide of civilisation.

As the Baliem Valley and its surrounding regions transform with the impact of change, we of the modern world have the unique opportunity not only to observe the stone age cultures grapple with the evolution of a new era, but to assist in giving advice, comfort and understanding to its traditional people.

Opposite: Now that the Baliem Valley houses many outsiders the Dani tribesmen build their villages and plant their sweet potato gardens on the lower slopes of the south western mountains above the valley. Walking from their homes with loaded headnets of produce for sale at the Wamena market Dani women and their children pause before descending to the valley floor.

Opposite: Dani watch towers, once used to house village sentries are now built as an entrance to a rest area on the road to Kurima.

A traditional Dani market on the outskirts of Wamena.

Opposite: On the morning of August 1999 the old market of twenty years and more lay in rubble. Burnt and dismantled the colourful Dani market scene will now only be found in guide books and memories.

Opposite: The newly constructed Wamena market lies on the road to Jiwika in the south western part of town. Housing many small warungs (shops) behind the market the area is set to become Wamena's main shopping centre.

Not too long ago the only way to Kurima in the eastern part of the Baliem Valley was through fields of unmarked territory. Only a single footpath led to this area. Today a wide semi-paved road sets the stage for speeding trucks and motor bikes bringing Kurima ever closer to Wamena and beyond.

Bibliography

Heider, Karl G., *The Dugum Dani*, Werner-Green Foundation for Anthropological Research Inc., NY, USA, or Viking Fund Publication in Anthropology, 1970.

Heider, Karl, G., *Grand Valley Dani, Peaceful Warriors*, Holt Rinehart & Wilson Inc., New York, USA, 1979.

Matthiessen, Peter, *Under the Mountain Wall*, Viking Press, USA,1962.

Koch, Klaus Friedrich, *War and Peace in Jalemo*, Harvard University Press, Massachusetts, USA, 1974.

Trenkenschuh, Frank A. and Msgr. Sowada, Alphonse, eds, *Asmat Sketch Books*, Advantage Printers, Chicago, USA, 1982.

Whitmore, T.C., *Tropical Rain Forest of the Far East*, Clarendon Press, Oxford, England, 1984.

Veevers-Carter, W., *Riches of the Rainforest*, Oxford University Press, Singapore, 1984.

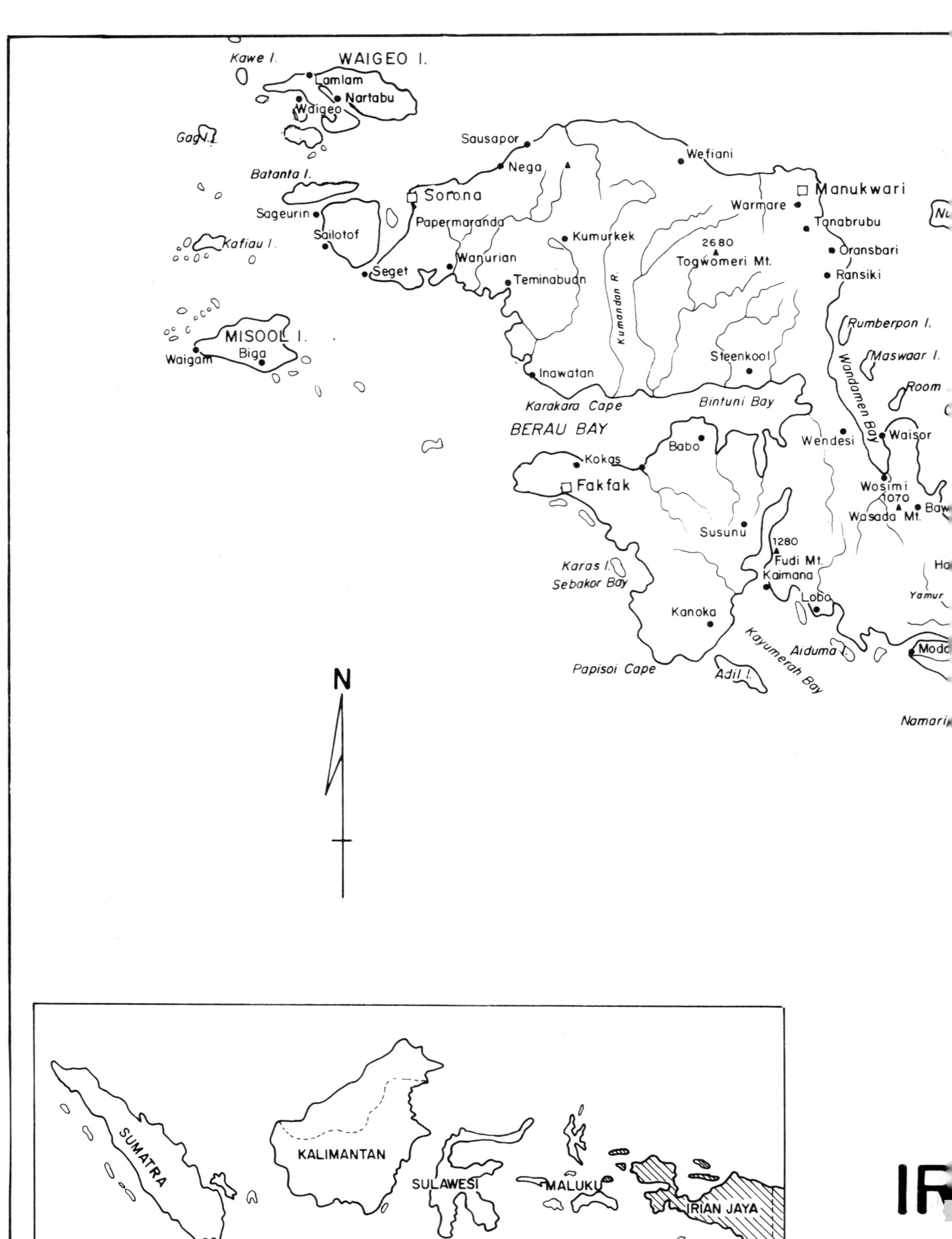

FEBRUARY, 1990